Explorations in Chemical Ecology

Edited by Thomas Eisner
and Jerrold Meinwald

*Invertebrate-Microbial Interactions: Ingested Fungal
Enzymes in Arthropod Biology*
by Michael M. Martin

Ecological Roles of Marine Natural Products
edited by Valerie J. Paul

ECOLOGICAL ROLES OF MARINE NATURAL PRODUCTS

Edited by

Valerie J. Paul

Marine Laboratory
University of Guam

Comstock Publishing Associates

a division of Cornell University Press

ITHACA AND LONDON

Library of Congress Cataloging-in-Publication Data

Ecological roles of marine natural products / Valerie J. Paul, editor.
    p. cm. — (Explorations in chemical ecology)
    Includes bibliographical references and index.
    ISBN 0-8014-2727-4 (alk. paper)
    1. Marine chemical ecology. 2. Marine metabolites—Physiological
effect. 3. Marine metabolites—Environmental aspects. I. Paul,
Valerie J. (Valerie Jean), 1957–   . II. Series.
QH541.5.S3E22   1992
574.5'2636—dc20   91-57899

Contents

Foreword

For some sixty years organic chemists have been interested in the diverse and novel structures of the secondary metabolites produced by marine organisms. Not surprisingly, many of these compounds have biological activities worthy of deeper study in connection with possible medicinal or agricultural applications. Separation techniques such as capillary gas chromatography and high-performance liquid chromatography have greatly simplified the isolation of pure chemical entities from complex mixtures, while X-ray crystallography, two-dimensional nuclear magnetic resonance spectroscopy, and mass spectrometry have advanced the art of structure determination beyond the wildest dreams of our scientific predecessors. Over the past decade, these developments have led to an explosive increase in our knowledge of the chemistry used by marine organisms.

As elegant and intriguing as the chemistry of natural products may be, however, it is but one essential element contributing to our understanding of marine chemical ecology. The ultimate aim of this discipline is to elucidate the chemical interactions that underlie such fundamental biological activities as food localization, predator deterrence, alarm communication, and promotion of sexual and reproductive behavior. Toward this end, biological and chemical studies need to go hand in hand. This goal is more easily set than attained.

To begin with, the natural histories of marine organisms such as brown algae or sea fans are intrinsically harder to study than those of their terrestrial counterparts. Field studies in marine environments are conceptually, technically, and financially demanding. In addition, bringing marine organisms into the laboratory is often far from simple. Nevertheless, there have been great advances in recent years, and ecological studies of marine environments in conjunction with sophisticated chemical approaches are no longer uncommon. The challenges in this field remain formidable, but there is no doubt that the opportunities for discovery are abundant; marine chemical ecology is certainly one of the great scientific frontiers of our time.

Valerie Paul has been a major contributor to this exciting area of research. For this volume, she has persuaded an outstanding group of authors to join her in presenting a series of authoritative accounts of some of the most interesting contemporary research in this rapidly evolving field. Her book should prove stimulating to biologists and chemists alike. Anyone seeking an introduction to marine chemical ecology will find an informed account of its current status. Those already in the field may appreciate the overview this volume offers on a wide variety of topics. Those of us who are otherwise occupied can enjoy *Ecological Roles of Marine Natural Products* for the opportunity it affords to savor the breadth and complexity of a fascinating, young science.

JERROLD MEINWALD
THOMAS EISNER

Ithaca, New York

Preface

Interest in the chemical ecology of marine organisms has grown tremendously over the past few years, and marine ecologists and natural products chemists have collaborated to produce some excellent research. This volume, one of the first to be devoted solely to discussions of the chemical ecology of marine organisms, reviews the recent research and discusses future directions for this rapidly expanding field. I hope it will provide both a background for marine biologists who are interested in entering this field and an ecological perspective for marine natural products chemists.

Because numerous reviews have been published covering the thousands of secondary metabolites isolated from marine organisms, the chemistry of marine natural products is not emphasized here. Instead, the book focuses on research that elucidates some of the ecological functions of marine secondary metabolites. I have enlisted contributions by several researchers who are actively engaged in studies of the chemical ecology of marine organisms. Each was asked to write about his own area of expertise and to emphasize the ecology and evolution of interactions mediated by marine secondary metabolites. Much of the research discussed here has been conducted within the past five years, and many references are made to articles that are in press or in preparation simply because this area of research is so new.

Each chapter attempts to integrate a limited amount of data into an evolutionary framework and to provide ideas for future research. Marine chemical ecology is in its infancy, and only limited research is available on which to base new hypotheses and research ideas. I hope that these hypotheses and viewpoints, whether ultimately proven right or wrong, will promote a great deal of research by marine biologists and chemists in chemical ecology over the next few years.

I thank the contributors to this book for their time and patience in seeing this project through to completion. Jerrold Meinwald and Thomas Eisner provided helpful comments on all chapters and a tremendous amount of support for this book. Finally, special thanks to William Fenical for sparking my interest in marine chemical ecology over a dozen years ago.

VALERIE J. PAUL

Mangilao, Guam

Contributors

D. John Faulkner, Scripps Institution of Oceanography, A-012F, University of California at San Diego, La Jolla, California 92093

Mark E. Hay, Institute of Marine Sciences, University of North Carolina at Chapel Hill, Morehead City, North Carolina 28557

Valerie J. Paul, Marine Laboratory, University of Guam, UOG Station, Mangilao, Guam 96923

Joseph R. Pawlik, Department of Biological Sciences and Center for Marine Science Research, University of North Carolina at Wilmington, Wilmington, North Carolina 28403-3297

Peter D. Steinberg, School of Biological Sciences, University of New South Wales, Kensington, New South Wales 2033, Australia

ECOLOGICAL ROLES OF MARINE NATURAL PRODUCTS

Introduction

VALERIE J. PAUL

Chemical ecology, or ecological biochemistry, examines the roles of naturally occurring compounds in plant and animal interactions. By definition, the field is interdisciplinary in scope and includes both chemical and biological research. Chemical investigations of the natural products produced by organisms include studies of structure, biosynthesis, organic synthesis, and mechanisms of action, while biological studies focus on the behavioral and developmental responses to chemical signals and the ecological consequences of these phenomena.

The chemistry of terrestrial plants and insects has been studied for the past century, and tens of thousands of different natural products have been isolated and chemically defined. This chemical knowledge of terrestrial organisms has contributed greatly to the development of the field of chemical ecology over the past few decades. Chemists have realized that the molecules they isolate and characterize often have potent biological activities and have likely evolved for specific biological functions. Biologists and ecologists have realized that chemical substances, particularly the secondary metabolites such as alkaloids, terpenoids, acetogenins, and aromatics, play an important role in complex behavioral and ecological interactions among organisms. The field has advanced most rapidly as the result of collaboration among chemists and biologists, including the incorporation of results and ideas from chemical research into biological research, and vice versa.

1

Studies on terrestrial organisms have contributed a great deal to our knowledge of the role of secondary metabolites as toxins, anti-feedants, pheromones, and allelopathic agents. Over the past fifteen years, chemical ecologists have proposed hypotheses regarding the evolution of chemical defenses in plants and animals (e.g., Feeny 1976; Rhoades and Cates 1976; Harborne 1978, 1982, 1984; Rhoades 1979; Fox 1981; Crawley 1983; Denno and McClure 1983; Spencer 1988). Many of these hypotheses suggest that the evolution of plant defense mechanisms is responsive to factors such as the plant's risk of discovery by herbivores, the cost of defense, and the relative value of various plant parts. These predictions are based on preliminary data, and in many cases have not been rigorously tested. Thus, although there is currently a general acceptance of the defensive roles of many secondary metabolites, there is considerable speculation regarding how herbivores and the physical environment interact to affect plant chemistry (Bryant et al. 1983, 1989; Coley 1983; Coley et al. 1985; Rhoades 1985). The diversity and ubiquity of secondary metabolites produced by plants have generated debate regarding their costs and benefits and the selective forces influencing their biosynthesis.

Research in terrestrial chemical ecology has provided a great deal of basic information that has advanced the fields of organic chemistry, biochemistry, ecology, behavior, and evolution. In addition, many practical applications have also developed. Knowledge of plant-insect interactions mediated by defensive compounds and other secondary metabolites has led to applications in the control of insect pests and microbial diseases in crop plants. Much of the pharmaceutical industry is based on terrestrial natural products or compounds modeled after these natural products.

Several thousand marine natural products have also been chemically defined; many of these are biologically active compounds possessing novel functional groups and molecular structures (Faulkner 1977, 1984a, 1984b, 1986, 1987, 1988, 1990, 1991; Scheuer 1978–1983, 1990). Interest in biotechnological applications for marine natural products has increased over the past decade as our knowledge of the chemistry of marine organisms has developed. Applied studies identifying future applications of these compounds as agrochemicals and pharmaceuticals have begun. Preliminary evidence suggests that marine organisms provide an untapped resource for future biotechnological applications. Several marine natural products are in clinical trials as anticancer and anti-inflammatory drugs (Ireland et al. 1988; Scheuer 1990). In contrast to terrestrial studies, however, much less is known about the natural functions of these metabolites in the marine environment. Only within the past few years have experimental evaluations of the role of marine natural products in the lives of the organisms that produce them been conducted.

In many ways, this is an opportune time for studies in marine chemical ecology, and the field is developing rapidly. Groups of marine plants and animals that are especially rich in secondary metabolites have been identified. Chemotaxonomic associations among these groups of organisms are being evaluated (e.g., Bergquist and Wells 1983). Our understanding of the natural history and ecology of marine organisms and the complexities of marine communities is advancing rapidly, and this facilitates investigations of how chemical interactions affect population and community structure. Results of field studies and manipulative ecological studies in many temperate and tropical intertidal and benthic subtidal communities have provided information on marine communities and the processes that structure them. Information on predator-prey interactions and competition in marine communities provides a basis for examining the role of secondary metabolites in mediating some of these ecological interactions. Unfortunately, in many marine habitats (including polar waters, deep-sea habitats, and the complex assemblages of coral reefs) lack of information about the inhabitants' natural history, behavior, and taxonomy complicates investigation and interpretation of the functions of secondary metabolites.

It is heartening to note that marine chemists and ecologists are becoming increasingly interested in collaborative studies. In fact, a new generation of marine chemical ecologists is developing. These individuals incorporate advances from both marine biology and natural products chemistry into their research.

In this book, I first provide a brief overview of the development of the field of marine chemical ecology. This includes a review of research in marine natural products chemistry and marine chemical ecology. Subsequent chapters contributed by investigators active in the field of marine chemical ecology review current research involving plant-herbivore and other predator-prey interactions. Current studies also include investigations of the role of marine natural products in antifouling, antimicrobial, and competitive interactions; in mediating larval settling and metamorphosis; and as pheromones. The emphasis of these chapters is biological, and the contributors have emphasized the ecological and evolutionary aspects of research on marine secondary metabolites.

Chemical Background

Over the past few decades, marine natural products have been isolated from many types of marine organisms, including seaweeds, sponges, coelenterates, echinoderms, and ascidians. The invertebrate phyla Porifera (sponges), Cnidaria (coelenterates), and Echinodermata (sea urchins and sea stars) are exclusively aquatic and principally marine. Since many of the organisms studied by marine natural products chemists are unique to the marine environment, it is not surprising that many novel secondary metabolites have been discovered.

Obvious analogies in the functions of marine and terrestrial natural products as toxins, antipredator defenses, and pheromones exist. The organisms found in marine and terrestrial habitats are different, however, and the two environments have evolved independently. Mechanisms of chemoreception and storing and exuding metabolites are influenced by the aqueous medium (see also Scheuer 1990) because diffusion of metabolites in water differs significantly from diffusion through air (Wilson 1970). Many marine organisms have planktonic larval stages that develop before the adults settle in near-shore communities. Following a period of vulnerability to pelagic predators and ocean currents, these organisms must develop techniques to locate suitable habitats for settlement. Thus, both the chemistry of marine natural products and the biology of some interactions mediated by these natural products can be expected to show some novel aspects. Chemical investigations of marine natural products have already highlighted the unique carbon skeletons and functional groups found in many marine secondary metabolites.

Recent studies of natural products produced by marine organisms have been reviewed by Faulkner (1984a, 1984b, 1986, 1987, 1988, 1990, 1991). These and other reviews of marine natural products (Faulkner 1977; Scheuer 1978–1983, 1987–1989; Fenical 1982, 1986; Okuda et al. 1986; Ireland et al. 1988) provide a thorough background of the numerous secondary metabolites that have been isolated. Bakus et al. (1986) discussed the classes of compounds isolated from marine organisms in relation to studies of marine chemical ecology. The structures and pharmacology of some marine toxins have been recently reviewed (Hall and Strichartz 1990). Marine organisms studied so far by natural products chemists include prokaryotic cyanobacteria and other microorganisms; phytoplankton such as dinoflagellates; the red, green, and brown macroalgae; sponges; coelenterates, including soft corals and gorgonians (Octocorallia), zoanthids, and stony corals; molluscs, especially the opisthobranchs; echinoderms, including the sea urchins (Echinoidea), sea stars (Asteroidea), and sea cucumbers (Holo-thuroidea); ascidians (Hemichordata); and fishes. The many classes of compounds that have been isolated include terpenoids, aromatics, acetogenins, saponins, nitrogenous compounds (including alkaloids), and peptides.

The seaweeds, opisthobranch molluscs, and sessile invertebrates (sponges, soft corals, gorgonian corals, ascidians) are the groups best studied by natural products chemists, so a good chemical background exists for them. They are the focus of this book, although not all are abundant or conspicuous members of marine communities. For exam-ple, some opisthobranch molluscs such as ascoglossans and nudi-branchs are common seasonally and only on particular host species, and they may be relatively inconspicuous in their natural habitats. Other invertebrates, such as sponges, soft corals, and gorgonian corals, can be abundant in terms of both numbers and biomass in some habi-tats, especially coral reefs.

Several major differences should be noted between the types and structures of secondary metabolites produced by marine and terres-trial organisms. Although alkaloids are common secondary metabolites in terrestrial plants (Robinson 1979), they are rare in marine plants. Terpenoids, acetogenins, and compounds of mixed biosynthesis are the major classes of secondary metabolites found in marine organ-isms. These compounds generally do not contain nitrogen. Many ma-rine environments, including the open ocean and the tropics, are ni-trogen poor, and the incorporation of nitrogen into secondary metabolites may not be metabolically feasible for most organisms liv-ing in these environments. The exception to this pattern is found in some phytoplankton, bacteria, and filamentous blue-green algae (cya-nobacteria), and in some invertebrates such as sponges, tunicates, and bryozoans that host bacteria or cyanobacteria in their tissues. Some of these microorganisms can fix elemental nitrogen, and therefore symbionts may provide nitrogen to their invertebrate hosts for bio-synthesis.

Many marine organisms incorporate halides from seawater into the organic compounds they produce. This biosynthetic feature is partic-ularly common in the red seaweeds, a few green seaweeds, and sponges (Fenical 1975). Red seaweeds from the families Bonnemaisoni-aceae, Rhizophyllidaceae, and Rhodomelaceae are rich in halogenated compounds that range in structure from halogenated methanes and halo-ketones to more complex terpenoids (Fenical 1975, 1982; Faulkner 1984a). The red algal genus *Laurencia*, the subject of extensive studies, produces hundreds of halogenated terpenoids and acetogenins representing a minimum of 26 structural classes. At least 16 of these classes are unknown in terrestrial organisms (Fenical 1975; Erickson 1983). Bromine is the most common halide found in marine secondary metabolites, but chlorine is also common. Halogenating enzymes such as bromoperoxidases and chloroperoxidases are proposed to function in the biosynthesis of these compounds, and these types of enzymes have been isolated (Fenical 1975; Barrow 1983; Neidleman and Geigert 1986). By contrast, terrestrial natural products rarely contain bromine, although some chlorinated compounds are known from fungi and higher plants.

Polymerized phenolic compounds known as tannins are abundant in higher plants and result from shikimic acid biosynthesis. Both condensed tannins (including lignans) and soluble tannins are known. Lignans are large polymers bound to polysaccharides in plant cell walls. Soluble or hydrolyzable tannins are formed from polymerization of esters of glucose with gallic acid or related compounds (Swain 1979). None of these types of tannins occur in marine plants. Polyphenolics occur in many of the brown seaweeds, but they result from a different biosynthetic pathway—polymerization of phloroglucinol (trihydroxybenzene) without sugar components. These metabolites are termed "phlorotannins" to distinguish them from the terrestrial tannins (Glombitza 1977; Ragan and Glombitza 1986).

Development of Marine Chemical Ecology

Undoubtedly, interest in marine chemical ecology developed as knowledge of the numerous and diverse natural products found in marine organisms increased. Marine natural products chemists began to ask questions about how the compounds they isolated function in nature. Early experiments addressed the antibiotic, antifungal, and ichthyotoxic effects of these metabolites, but not all used marine microorganisms or marine fish to test the hypotheses that these compounds were toxic and deterrent to predators and pathogens. Experiments are now being more carefully designed and replicated for statistical analysis, and naturally co-occurring predatory fishes and pathogenic microorganisms are being used to test the deterrent effects of compounds from marine organisms. Progress in experimental design and analysis of these types of bioassays has been made in the past five years, especially as marine ecologists have become more interested in chemical ecology (e.g., Peterson and Renaud 1989).

Manipulative field studies are now common in both terrestrial and marine ecology. In many experiments, competitors or predators are either added to or excluded from a habitat to understand their influence on community structure. Similarly, secondary metabolites can be added to artificial diets and tested as feeding deterrents against natural herbivores or predators in field studies (Harvell et al. 1988; reviewed by Hay and Fenical 1988).

Terrestrial chemical ecology has served as a model for many of the studies in marine chemical ecology. Research in terrestrial chemical ecology has focused on animal-plant interactions and has resulted in many papers and review articles on "theories" of plant antiherbivore chemistry and biochemical evolution. Models of both intraspecific and interspecific patterns of variation in the production of secondary metabolites by plants have been proposed.

Two models that describe intraspecific patterns of secondary metabolite allocation are optimal defense theory and the carbon-nutrient theory. Optimal defense theory suggests that chemical defenses are costly, and plants allocate secondary metabolites to parts of the plant that are most valuable to the plant or most susceptible to herbivores (McKey 1979; Rhoades 1979; Nitao and Zangerl 1987). Allocation of defensive compounds within plants thus reflects the value of different plant tissues in terms of their relative contributions to fitness. In the carbon-nutrient theory, resource availability affects the phenotypic expression of chemical defenses. Allocation of resources to chemical defenses can change as environmental conditions such as light or nutrient availability change. When resources exceed the necessary levels for plant growth, the excess is shunted to chemical defense. For example, when nutrient levels are low and restrict growth, increases in light or CO_2 levels result in excess carbon being used for production of carbon-based secondary metabolites. The predictions of the model therefore depend on the relative availability of carbon and nitrogen to the plants (Bryant et al. 1983, 1988, 1989).

Two important models involving interspecific variation in plant chemical defenses have developed: the plant apparency model and the resource availability model. The models are similar to those described for intraspecific variation. In the plant apparency model, hypotheses focus on the predictability (apparency) of plants—and therefore their likelihood of being detected by herbivores—in relation to their allocation of chemical defenses (Feeny 1976; Rhoades and Cates 1976; Rhoades 1979; Cates 1980). The model proposes that plant defense mechanisms have evolved in response to the plant's risk of discovery by herbivores, the cost of defense, and the value of various plant parts. Both the types and the amounts of defenses evolved by plants are considered to be directly related to the risk of discovery of plants or plant tissues by herbivores (Feeny 1975, 1976; Rhoades and Cates 1976; Rhoades 1979). Ephemeral (unapparent) plants or plant tissues are at lower risk of herbivore attack and therefore allocate less energy to chemical defenses. Plants or tissues that are predictable and available (apparent) are at greater risk and will allocate more energy to the production of secondary metabolites.

Coley et al. (1985) developed an alternative hypothesis, one that involves plants' allocation of resources to chemical defenses based on resource availability in the environment. They suggest that species growing in nutrient-poor or low-light environments will have inherently slow growth rates. Because the tissues of these plants are produced slowly, they are more valuable to the plants and will be more heavily defended. Plants with fast growth rates will produce lower amounts of secondary metabolites.

The plant apparency and resource availability hypotheses are not mutually exclusive, and both fit many observed patterns of chemical defenses found among terrestrial plants. Both also distinguish among types of defenses produced by plants in high versus low concentrations and the costs associated with their production. Compounds such as terpenoids and alkaloids are usually produced in low concentrations by plants (<5% and usually only 1–2% of dry weight) and are termed "toxins" or "qualitative defenses." Compounds such as tannins and lignins are produced in much higher concentrations (often 10–20% or more of plant dry weight) and are termed "digestibility reducers" or "quantitative defenses." Quantitative defenses are assumed to be more costly to plants because they are produced in such high concentrations. They are also proposed as more effective deterrents against a broad array of herbivores because they bind digestive enzymes, inhibit the basic digestive processes, and allow little possibility for counteradaptation by herbivores.

Many of the basic assumptions of these models have not been carefully evaluated experimentally. As evidence accumulates, it appears that the proposed distinctions in costs and functions of quantitative versus qualitative defenses may not be warranted (see also Hay and Fenical 1988; Fagerström 1989). Both types of defenses work in a dosage-dependant manner; higher concentrations of terpenoids or tannins are usually better deterrents against nonadapted herbivores than lower concentrations. Not enough is known about the synthesis, storage, turnover, and release of secondary metabolites to determine whether tannins are more costly than terpenoids or alkaloids. Both qualitative and quantitative defenses can be sequestered or exuded; thus turnover rates and costs could vary widely for both types of defenses. Many insects and mammalian grazers eat plants high in tannins; therefore, counteradaptation to these defenses can clearly occur (Bernays et al. 1989). Continued investigations into the biosynthesis and turnover of secondary metabolites and interactions between chemical defenses and predators and pathogens in terrestrial and marine systems will provide data on which to evaluate these hypotheses (see Southwood et al. 1986; Bryant et al. 1989; Baldwin et al. 1990). Additionally, the physiological constraints on secondary metabolite production suggested by the carbon-nutrient and resource availability hypotheses will be an important research direction for marine chemical ecologists.

It is particularly interesting to determine how these models fit our observations on marine systems. Several contributors to this book make comparisons between terrestrial and marine animal-plant interactions based on these predictions (see also Hay 1991). Marine predator-prey interactions can be examined similarly because many sessile invertebrates produce large amounts of secondary metabolites. Thus, marine chemical ecologists are now in a position to contribute to the development of hypotheses and models of the evolution of chemical defenses and other chemical interactions among organisms.

Several natural history patterns differ markedly between marine and terrestrial habitats and may have important implications for studies of chemical ecology. First, specialists appear to be much more common in terrestrial than in marine habitats. Most herbivorous insects are specialized for feeding on just one or a few related plants (Bernays and Graham 1988). Mammalian herbivores such as the koala and panda also have highly specific diets. The natural history and feeding patterns of many marine herbivores and predators, especially the mesograzers such as amphipods, crabs, and molluscs, are not known; however, present information suggests that few of these organisms are specialists with a very narrow host range (see Hay and Fenical 1988; Hay, this volume). Dominant herbivorous species such as sea urchins, gastropod snails, and fishes are generalists with broad diets. The best known examples of specialists in the marine environment are the opisthobranch molluscs, including nudibranchs, which specialize on particular species of sponges, coelenterates, or tunicates, and the ascoglossans, which specialize on particular seaweeds.

Pathogens of many terrestrial plants are specialized for one or a few related plants, and it is possible that they are more specialized than marine pathogens. Information on plant diseases is especially well known for crop plants. By contrast, relatively little is known about pathogens of marine plants (Andrews 1977; Goff and Glasgow 1980) and animals (Peters 1988; Leibovitz and Koulish 1989). Recent examples of marine pathogens may include black-band disease in corals (Peters 1988), wasting disease in the eelgrass *Zostera marina* (Muehlstein et al. 1988), and an unidentified viral epidemic that killed the sea urchin *Diadema* in the Caribbean (Lessios et al. 1984; Lessios 1988). These appear to be host specific, although black-band disease can affect many species of stony corals and gorgonian corals (reviewed in Peters 1988). How marine secondary metabolites mediate host-pathogen interactions has not been investigated.

A second difference between the natural history patterns of marine organisms and those of terrestrial organisms is the mechanisms of larval dispersal. The mechanisms used in terrestrial habitats have likely contributed to a greater degree of host specialization in terrestrial herbivores. In the case of many specialized insects, highly mobile adults search out the appropriate host plant and oviposit on it. Consequently, less mobile larval forms that may be subject to intense predation pressures do not have to try to locate specific food items. By contrast, many marine animals have a planktonic larval stage that may be dispersed by ocean currents. Larvae are susceptible to predation while in the plankton as well as when they return to near-shore communities to settle (Gaines and Roughgarden 1987; Roughgarden et al. 1988). Predation pressures may restrict the time larvae spend searching for specific hosts, particularly if these hosts are rare. As exceptions, some ascoglossans and nudibranchs have direct-development larvae, and amphipods brood their eggs. Based on our current knowledge, these organisms are among the major examples of specialization in marine habitats (Hay et al. 1987; Hay and Fenical 1988; Hay, this volume).

Overview of the Chapters

The first three chapters in this book discuss plant-herbivore interactions in the marine environment. Some of the earliest studies in marine chemical ecology dealt with these interactions and were often patterned after terrestrial studies. In Chapter 1, I present research on chemical defenses of seaweeds on coral reefs. Tropical seaweeds have yielded the majority of the approximately 700 seaweed metabolites that have been isolated and chemically defined, including numerous compounds that have shown toxicity and feeding deterrent activity in many different field and laboratory experiments (Paul and Fenical 1987). It is generally believed that tropical seaweeds produce a large variety of secondary metabolites because herbivory is intense on coral reefs. I discuss the responses of different tropical herbivores to seaweed chemical defenses and the variation in concentrations and types of metabolites in tropical seaweeds.

Chapter 2, written by Peter Steinberg, considers the types of defensive chemicals produced by brown seaweeds, primarily terpenes and polyphenolics. Polyphenolics are produced by brown seaweeds of the orders Fucales and Laminariales and can occur in high concentrations (5–15% dry mass) in these plants. The chapter focuses on the distribution of secondary metabolites within different taxa of the brown seaweeds, the biological activity of the polyphenolics toward herbivores, and the biogeographical distribution of the seaweeds containing polyphenolics. Although concentrations of polyphenolics are high within temperate brown seaweeds, they are absent or in low concentrations (0–1% dry mass) in most of the tropical brown algae studied so far. Terpenes and other nonpolar metabolites are more common in tropical brown seasweeds. Comparisons are made between the polyphenolics in seaweeds and their effects on marine herbivores and terrestrial plant tannins and their effects on herbivores.

Chapter 3, written by Mark Hay, deals with the role of secondary metabolites in mediating feeding specialization by some marine herbivores. Many of the specialized herbivores are small grazers such as amphipods, crabs, polychaetes, and opisthobranch molluscs. Hay hypothesizes that these small sedentary herbivores are severely affected by predation and therefore live and feed on chemically defended seaweeds that provide a refuge from predators. He also discusses other complex interactions such as associations between defended and undefended seaweeds and the effects of seaweed secondary metabolites on carbon and nutrient cycling.

John Faulkner's contribution on chemical defenses of marine molluscs (Chapter 4) complements Hay's review in Chapter 3, since it discusses molluscs such as nudibranchs and ascoglossans that are specialized to feed on just one or a few host species. Faulkner and Ghiselin (1983) hypothesized that in the evolution of the opisthobranch molluscs, loss of the shell as a physical defense was related to the ability of these animals to sequester the host's chemical defenses and utilize them against predators. The relationships between host organisms, opisthobranchs, and higher-order predators are often complex (Rogers and Paul, 1991), and the defensive chemistry of the opisthobranchs and their hosts likely plays an important role in these interactions. Although a great deal is known about the chemistry of these molluscs, much less is known about their chemical ecology, and only a few studies have experimentally evaluated the function of sequestered metabolites of opisthobranchs as chemical defenses. This chapter provides an excellent review of research in this area and highlights areas for future research.

The final chapters examine the chemical ecology of sessile marine invertebrates. In Chapter 5 I examine the chemical defenses of sessile invertebrates and emphasize studies of feeding deterrents and antifouling compounds. Chemical defenses of marine invertebrates have not been studied as well as those of seaweeds, but evidence supporting the deterrent role of many compounds is accumulating rapidly. The natural products produced by sponges, coelenterates, and ascidians are structurally very different, ranging from terpenoids to alkaloids, yet their natural functions may be very similar.

Chapter 6, written by Joseph Pawlik, discusses settlement processes of benthic invertebrates and evidence for a role for secondary metabolites in these processes. Since settlement of marine invertebrates is critical to the structure of benthic communities, understanding settlement cues and processes will contribute significantly to marine ecology. Many studies have provided evidence that chemical cues stimulate or inhibit settlement of marine invertebrate larvae; however, few have resulted in the isolation and characterization of active compounds. This chapter considers those studies that provide a chemical basis for settlement processes.

All the chapters focus on research with a strong chemical basis. Often, secondary metabolites were actually isolated and chemically defined, and the compounds were found to be responsible for biological activities such as feeding deterrence, settlement induction, or antifouling activity. At the same time, the contributors emphasize the importance of using relevant bioassays to test for biological activities. Ideally, compounds should be tested against several different naturally co-occurring predators or fouling organisms to determine how specific or broadly active they are. The use of field assays to test for bioactivities can provide information on how natural products act with regard to natural populations of predators or fouling organisms.

It is important to emphasize the possible multifunctionality of marine secondary metabolites. Compounds may have several functions in addition to being herbivore or predator deterrents (see Krischik et al. 1991). Marine secondary metabolites may play a role as deterrents against pathogenic microorganisms, fouling organisms, or competitors (Bakus et al. 1986; Paul and Fenical 1986; Coll et al. 1987; Gerhart et al. 1988; Sammarco and Coll 1988; Davis et al. 1989; de Nys et al. 1991), and as pheromones (Pawlik, this volume) and reproductive cues (Pass et al. 1989). It is likely that secondary metabolites play multiple adaptive roles for the producing organisms and therefore may not have evolved primarily to serve a defensive function (Sammarco and Coll 1990).

Other Research in Marine Chemical Ecology

Several other areas of marine natural products chemistry have begun to interest chemical ecologists, although they have not been as well studied as the seaweed-herbivore and benthic invertebrate-predator interactions. Except for the following brief discussion, these areas of research are not emphasized in other chapters of this book.

Phytoplankton

Marine phytoplankton (microalgae) are known to produce toxins that affect humans (Okaichi et al. 1989), and natural products chemists have studied these well. For instance, saxitoxin (1)

1 X = H
2 X = OH

and related metabolites are the causative toxins involved in paralytic shellfish poisoning (PSP), which affects thousands of people each year (Hashimoto 1979; Russell 1984; Steidinger and Baden 1984). The toxins are concentrated in shellfish that feed on *Gonyaulax tamarensis* and other dinoflagellates that produce these toxic metabolites. The chemistry and biological significance of the compounds saxitoxin, neosaxitoxin (2), and related toxins are reviewed by Shimizu (1984), who has worked extensively with the chemistry of toxic dinoflagellates. Reviews of the chemistry and pharmacology of these toxins are presented by Hall and Strichartz (1990).

Some predators on shellfish, such as sea otters, may be able to detect high levels of saxitoxin in the butter clams they eat and may avoid consuming lethal amounts of toxin by discarding the most toxic body parts and reducing consumption rates (Kvitek et al. 1991). Saxitoxin may be an acquired chemical defense for the clams, since high

amounts are concentrated in the siphons, the structures most exposed to predators (Kvitek and Beitler 1991; Kvitek et al. 1991).

The dinoflagellate *Ptychodiscus brevis* (= *Gymnodinium breve*) is
responsible for the "red tides" of the southeastern Atlantic Coast and
the Gulf of Mexico that result in fish kills. The structure of the polyether brevetoxin B (3)

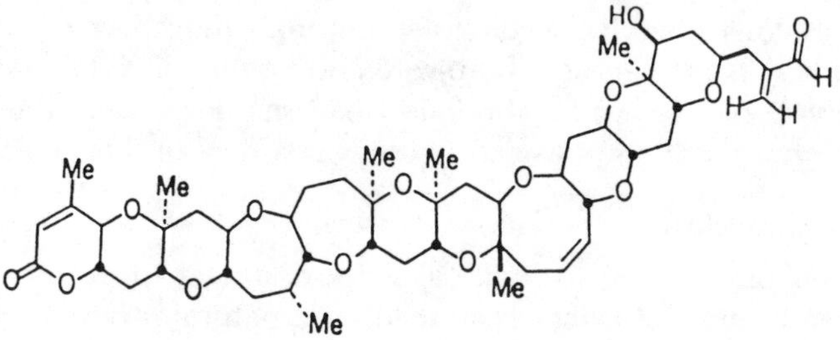

3

was determined by X-ray crystallography (Lin et al. 1981). Several poly-
cyclic ethers closely related to brevetoxin B are now known (Faulkner
1984a, 1986; Nakanishi 1988; Hall and Strichartz 1990).

Other dinoflagellate toxins include those responsible for diarrhetic
shellfish poisoning, produced by *Dinophysis fortii* (Yasumoto et al.
1985). The dinoflagellate *Prorocentrum minimum* and the diatom *Navicula delognei* f. *elliptica* both produce metabolites that show antimicrobial activity (Faulkner 1984a, 1986). The causative agent of ciguatera (fish poisoning), which may result in death, has been linked to
the tropical dinoflagellate *Gambierdiscus toxicus*. The structure of the
toxic polyether metabolite ciguatoxin (4)

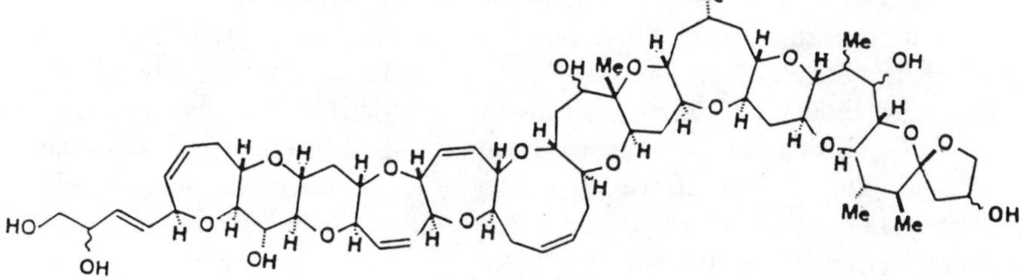

4

involved in ciguatera poisoning has recently been reported (Murata et
al. 1989).

Chemical investigations of marine phytoplankton have been complicated by the difficulty in growing them and by the low yields of their secondary metabolites under culture conditions. Many of the secondary metabolites are released into the culture medium and are only recognized when they have toxic properties. Thus, other compounds that might show feeding deterrent or allelopathic activity may not be recognized. These limitations also affect studies of the chemical ecology of these organisms. Studies have shown that planktonic herbivores such as copepods, as well as sessile, filter-feeding invertebrates such as mussels, have feeding preferences and avoid some species of phytoplankton and blue-green algae (Huntley et al. 1986; Van Alstyne 1986; Fulton and Paerl 1987; Ward and Targett 1989). Recently, Starr et al. (1990) suggested that spawning by some marine invertebrates such as sea urchins and mussels may be triggered by metabolites released into seawater by phytoplankton. This direct coupling of marine invertebrate spawning with phytoplankton blooms may be advantageous for larval growth and survival. A great deal of work remains to be done on the chemical ecology of phytoplankton natural products.

Echinoderms

The echinoderms include asteroids (sea stars), crinoids (sea lilies), holothuroids (sea cucumbers), echinoids (sea urchins), and ophiuroids (brittle stars). Saponins, a group of water-soluble isoprenoid glycosides, are the predominant secondary metabolites that have been isolated from echinoderms, especially from sea cucumbers and sea stars. Burnell and ApSimon (1983) provided an excellent review of echinoderm saponins, and Minale et al. (1982) reported on the methods used in saponin research. The saponins of the asteroids are primarily steroidal glycosides, whereas the holothurian saponins are based on triterpenoids rather than steroids. More recent studies of echinoderm saponins have been reviewed by Faulkner (1984b, 1986) and Stonik and Elyakov (1988).

Little is known about the chemical ecology of the sea stars and sea cucumbers that produce saponins. McClintock and Vernon (1990) suggested that the eggs and embryos of Antarctic sea stars are chemically defended, but they did not determine types of compounds involved. Some of the saponin-producing echinoderms have been reported to be toxic (Bakus 1974, 1981; Bakus and Green 1974; Mackie et at. 1975; McClintock 1989), and saponins have been demonstrated to be the unpalatable constituents of some echinoderms (Lucas et al. 1979; Bingham and Braithwaite 1986).

Fishes

Several fish species contain toxic compounds in their body parts or produce defensive secretions (Tachibana 1988). The potent neurotoxin tetrodotoxin (5)

5

has been isolated from livers and ovaries of pufferfishes (family Tetrodontidae). In Japan, pufferfish fillets are a delicacy and are specially prepared and consumed despite their sometimes fatal effect. Tetrodotoxin has also been isolated from several species of chaetognaths (arrowworms; Thuesen et al. 1988; Thuesen and Kogure 1989), the venom glands of the blue-ringed octopus (Sheumack et al. 1978), the Taiwan goby (*Gobius cringer*), several newts of the genus *Taricha*, the skin of the Costa Rican frog *Atelopus*, and the sea star *Astropecten* (Miyazawa et al. 1985, 1987; reviewed in Fenical 1986). Evidence now suggests that tetrodotoxin may be produced by bacteria associated with these organisms (Yasumoto et al. 1986; Tachibana 1988; Thuesen and Kogure 1989).

Flatfishes such as the Red Sea moses sole (*Pardachirus marmoratus*) produce defensive secretions that repel sharks. The chemical structures of the deterrent metabolites are not known from *P. marmoratus*, but the related Indo-Pacific sole *P. pavoninus* produces aminoglycoside saponins and peptides. The compounds have shark-repellent effects at low concentrations (Fenical 1986; Nakanishi 1988; Tachibana 1988).

Pheromones

Simple hydrocarbons produced by brown algae have been reported to function as sex pheromones. Female gametes produce these simple, volatile, usually C-11 hydrocarbons that attract the male gametes. Müller and his co-workers first elucidated the function of these hydrocarbons and have now reported the structures of many compounds from different species of brown algae (Moore 1978; Müller et al. 1979, 1981; Faulkner 1984a; Maier and Müller 1986).

Diterpenoids may play a role as sex pheromones in coelenterate species. Some of the compounds found in soft corals of the genus *Sinularia* are present only in eggs and may attract sperm to the eggs (Sammarco and Coll 1988; Coll et al. 1989). A similar function of sperm attraction has been suggested for lipid-derived molecules in stony corals (Coll et al. 1987).

Waterborne sperm attractants that are species specific have been reported from free-spawning starfish (Miller 1989); however, actual secondary metabolites have not been isolated. Clearly, a great deal of work with pheromones remains to be done with marine plants and animals.

References

Andrews, J.H. 1977. Observations on the pathology of seaweeds in the Pacific Northwest. Can. J. Bot. 55:1019–1027.
Bakus, G.J. 1974. Toxicity in holothurians: a geographic pattern. Biotropica 6:229–236.
Bakus, G.J. 1981. Chemical defense mechanisms and fish feeding behavior on the Great Barrier Reef, Australia. Science 211:497–499.
Bakus, G.J., and Green, G. 1974. Toxicity in sponges and holothurians: a geographic pattern. Science 185:951–953.
Bakus, G.J., Targett, N.M., and Schulte, B. 1986. Chemical ecology of marine organisms: an overview. J. Chem. Ecol. 12:951–987.
Baldwin, I.T., Sims, C.L., and Kean, S.E. 1990. The reproductive consequences associated with inducible alkaloid responses in wild tobacco. Ecology 71:252–262.
Barrow, K.D. 1983. Biosynthesis of marine metabolites. *In* Marine natural products: chemical and biological perspectives, vol. 5, ed. P.J. Scheuer, pp. 51–86. New York: Academic Press.
Bergquist, P.R., and Wells, R.J. 1983. Chemotaxonomy of the Porifera: the development and current status of the field. *In* Marine natural products: chemical and biological perspectives, vol. 5, ed. P.J. Scheuer, pp. 1–50. New York: Academic Press.
Bernays, E.A., Cooper Driver, G., and Bilgener, M. 1989. Herbivores and plant tannins. Adv. Ecol. Res. 19:263–302.
Bernays, E.A., and Graham, M. 1988. On the evolution of host specificity in phytophagous arthropods. Ecology 69:886–892.
Bingham, B.L., and Braithwaite, L.F. 1986. Defense adaptations of the dendrochirote holothurian *Psolus chitonoides* Clark. J. Exp. Mar. Biol. Ecol. 98:311–322.
Bryant, J.P., Chapin, F.S., III, and Klein, D.R. 1983. Carbon/nutrient balance of boreal plants in relation to vertebrate herbivory. Oikos 40:357–368.
Bryant, J.P., Kuropat, P.J., Cooper, S.M., Frisby, K., and Owen-Smith, N. 1989. Resource availability hypothesis of plant antiherbivore defence tested in a South African savanna ecosystem. Nature 340:227–229.

Bryant, J.P., Tuomi, J., and Niemala, P. 1988. Environmental constraint of constitutive and long-term inducible defenses in woody plants. *In* Chemical mediation of coevolution, ed. K.C. Spencer, pp. 367–390. San Diego: Academic Press.

Burnell, D.J., and ApSimon, J.W. 1983. Echinoderm saponins. *In* Marine natural products: chemical and biological perspectives, vol. 5, ed. P.J. Scheuer, pp. 287–389. New York: Academic Press.

Cates, R.G. 1980. Feeding patterns of monophagous, oligophagous, and polyphagous insect herbivores: the effect of resource abundance and plant chemistry. Oecologia 46:22–31.

Coley, P.D. 1983. Herbivory and defensive characteristics of tree species in a lowland tropical forest. Ecol. Monogr. 53:209–233.

Coley, P.D., Bryant, J.P., and Chapin, F.S., III. 1985. Resource availability and plant antiherbivore defense. Science 230:895–899.

Coll, J.C., Bowden, B.F., Heaton, A., and Clayton, M. 1987. Chemotaxis in tropical alcyonacean corals. *In* AMSA abstracts, Australian Marine Sciences Association Conference, vol. 1, p. 12. Townsville, Australia.

Coll, J.C., Bowden, B.F., Heaton, A., Scheuer, P.J., Li, M.K.W., Clardy, J., Schulte, G.K., and Finer-Moore, J. 1989. Structures and possible functions of pukalide and epoxypukalide: diterpenes associated with eggs of sinularian soft corals (Cnidaria, Anthozoa, Octocorallia, Alcyonacea, Alcyoniidae). J. Chem. Ecol. 15:1177–1191.

Coll, J.C., Price I.R., Konig, G.M., and Bowden, B.F. 1987. Algal overgrowth of alcyonacean soft corals. Mar. Biol. 96:129–135.

Crawley, M.J. 1983. Herbivory: the dynamics of animal-plant interactions. Berkeley: University of California Press.

Davis, A.R., Targett, N.M., McConnell, O.J., and Young, C.M. 1989. Epibiosis of marine algae and benthic invertebrates: natural products chemistry and other mechanisms inhibiting settlement and overgrowth. *In* Bioorganic marine chemistry, vol. 3, ed. P.J. Scheuer, pp. 85–114. Berlin: Springer-Verlag.

Denno, R.F., and McClure, M.S., eds. 1983. Variable plants and herbivores in natural and managed systems. New York: Academic Press.

de Nys, R., Coll, J.C., and Price, I.R. 1991. Chemically mediated interactions between the red alga *Plocamium hamatum* (Rhodophyta) and the octocoral *Sinularia cruciata* (Alcyonacea). Mar. Biol. 108:315–320.

Erickson, K.L. 1983. Constituents of *Laurencia*. *In* Marine natural products: chemical and biological perspectives, vol. 5, ed. P.J. Scheuer, pp. 132–257. New York: Academic Press.

Fagerström, T. 1989. Anti-herbivory chemical defense in plants: a note on the concept of cost. Am. Nat. 133:281–287.

Faulkner, D.J. 1977. Interesting aspects of marine natural products chemistry. Tetrahedron 33:1421–1443.

Faulkner, D.J. 1984a. Marine natural products: metabolites of marine algae and herbivorous marine molluscs. Nat. Prod. Rep. 1:251–280.

Faulkner, D.J. 1984b. Marine natural products: metabolites of marine invertebrates. Nat. Prod. Rep. 1:551–598.

Faulkner, D.J. 1986. Marine natural products. Nat. Prod. Rep. 3:1–33.

Faulkner, D.J. 1987. Marine natural products. Nat. Prod. Rep. 4:539–576.

Faulkner, D.J. 1988. Marine natural products. Nat. Prod. Rep. 5:613–663.

Faulkner, D.J. 1990. Marine natural products. Nat. Prod. Rep. 7:269–309.

Faulkner, D.J. 1991. Marine natural products. Nat. Prod. Rep. 8:97–147.

Faulkner, D.J., and Ghiselin, M.T. 1983. Chemical defense and evolutionary ecology of dorid nudibranchs and some other opisthobranch gastropods. Mar. Ecol. Prog. Ser. 13:295–301.

Feeny, P. 1975. Biochemical evolution between plants and their insect herbivores. *In* Coevolution of animals and plants, ed. L.E. Gilbert and P.H. Raven, pp. 1–19. Austin: University of Texas Press.

Feeny, P. 1976. Plant apparency and chemical defense. Recent Adv. Phytochem. 10:1–40.

Fenical, W. 1975. Halogenation in the Rhodophyta: a review. J. Phycol. 11:245–259.

Fenical, W. 1982. Natural products chemistry in the marine environment. Science 215:923–928.

Fenical, W. 1986. Marine alkaloids and related compounds. *In* Alkaloids: chemical and biological perspectives, vol. 4, ed. S.W. Pelletier, pp. 275–330. New York: John Wiley and Sons.

Fox, L.R. 1981. Defense and dynamics in plant-herbivore systems. Am. Zool. 21:853–864.

Fulton, R.S., III, and Paerl, H.W. 1987. Toxic and inhibitory effects of the blue-green alga *Microcystis aeruginosa* on herbivorous zooplankton. J. Plankton Res. 9:837–855.

Gaines, S., and Roughgarden, J. 1987. Fish in offshore kelp forests affect recruitment to intertidal barnacle populations. Science 235:479–481.

Gerhart, D.J., Rittschof, D.J., and Mayo, S.W. 1988. Chemical ecology and the search for marine antifoulants. J. Chem. Ecol. 14:1905–1917.

Glombitza, K.W. 1977. Highly hydroxylated phenols of the Phaeophyceae. *In* Marine natural products chemistry, ed. D.J. Faulkner and W. Fenical, pp. 191–204. New York: Plenum Press.

Goff, L.J., and Glasgow, J.C. 1980. Pathogens of marine plants. Special Publ. no. 7. Santa Cruz: Center for Coastal Marine Studies, University of California.

Hall, S., and Strichartz, G., ed. 1990. Marine toxins: origin, structure, and molecular pharmacology. Am. Chem. Soc. Symp. Ser. No. 418. Washington, D.C.: American Chemical Society.

Harborne, J.B., ed. 1978. Biochemical aspects of plant and animal coevolution. New York: Academic Press.

Harborne, J.B. 1982. Introduction to ecological biochemistry. New York: Academic Press.

Harborne, J.B. 1984. Recent advances in chemical ecology. Nat. Prod. Rep. 3:323–344.

Harvell, C.D., Fenical, W., and Greene, C.H. 1988. Chemical and structural defenses of Caribbean gorgonians (*Pseudopterogorgia* spp.). I. Development of an in situ feeding assay. Mar. Ecol. Prog. Ser. 49:287–294.

Hashimoto, Y. 1979. Marine toxins and other bioactive metabolites. Tokyo: Japan Scientific Society.

Hay, M.E. 1991. Marine-terrestrial contrasts in the ecology of plant chemical defenses against herbivores. Trends Ecol. Evol., in press.

Hay, M.E., Duffy, J.E., Pfister, C.A., and Fenical, W. 1987. Chemical defense against different marine herbivores: are amphipods insect equivalents? Ecology 68:1567–1580.

Hay, M.E., and Fenical, W. 1988. Chemically-mediated seaweed herbivore interactions. Annu. Rev. Ecol. Syst. 19:111–145.

Huntley, M., Sykes, P., Rohan, S., and Marin, V. 1986. Chemically mediated rejection of dinoflagellate prey by the copepods *Calanus pacificus* and *Paracalanus parvus*: mechanism, occurrence and significance. Mar. Ecol. Prog. Ser. 28:105–120.

Ireland, C.M., Roll, D.M., Molinski, T.F., McKee, T.C., Zabriskie, T.M., and Swersey, J.C. 1988. Uniqueness of the marine chemical environment: categories of marine natural products from invertebrates. *In* Biomedical importance of marine organisms, ed. D. Fautin, pp. 41–58. Mem. Cal. Acad. Sci. Symp. No. 13. San Francisco: California Academy of Sciences.

Krischik, V.A., Goth, R.W., and Barbosa, P. 1991. Generalized plant defense: effects on multiple species. Oecologia 85:562–571.

Kvitek, R.G., and Beitler, M.K. 1991. Relative insensitivity of butter clam neurons to saxitoxin: a pre-adaptation for sequestering paralytic shellfish poisoning toxins as a chemical defense. Mar. Ecol. Prog. Ser. 69:47–54.

Kvitek, R.G., DeGange, A.R., and Beitler, M.K. 1991. Paralytic shellfish poisoning toxins mediate feeding behavior of sea otters. Limnol. Oceanogr. 36:393–404.

Leibovitz, L., and Koulish, S. 1989. A viral disease of the ivory barnacle, *Balanus eburneus*, Gould (Crustacea, Cirripedia). Biol. Bull. 176:301–307.

Lessios, H.A. 1988. Mass mortality of *Diadema antillarum* in the Caribbean: what have we learned? Annu. Rev. Ecol. Syst. 19:371–393.

Lessios, H.A., Robertson, D.R., and Cubit, J.D. 1984. Spread of *Diadema* mass mortality through the Caribbean. Science 226:335–337.

Lin, Y.Y., Risk, M., Ray, S.M., Van Engen, D., Clardy, J., Golik, J., James, J.C., and Nakanishi, K. 1981. Isolation and structure of brevetoxin B from the "red tide" dinoflagellate *Ptychodiscus brevis* (*Gymnodinium breve*). J. Am. Chem. Soc. 103:6773–6775.

Lucas, J.S., Hart, R.J., Howden, M.E., and Salathe, R. 1979. Saponins in eggs and larvae of *Acanthaster planci* (L.) (Asteroidea) as chemical defences against planktivorous fish. J. Exp. Mar. Biol. Ecol. 40:155–165.

McClintock, J.B. 1989. Toxicity of shallow-water Antarctic enchinoderms. Polar Biol. 9:461–465.

McClintock, J.B., and Vernon, J.D. 1990. Chemical defense in the eggs and embryos of antarctic sea stars (Echinodermata). Mar. Biol. 105:491–495.

McKey, D. 1979. The distribution of secondary compounds within plants. *In* Herbivores: their interactions with secondary plant metabolites, ed. G.A. Rosenthal and D.H. Janzen, pp. 56–133. New York: Academic Press.

Mackie, A.M., Singh, H.T., and Fletcher, T.C. 1975. Studies on the cytolytic effects of seastar (*Marthasterias glacialis*) saponins and synthetic surfactants in the plaice *Pleuronectes platessa*. Mar. Biol. 29:307–314.

Maier, I., and Müller, D.G. 1986. Sexual pheromones in algae. Biol. Bull. 170:145–175.

Miller, R.L. 1989. Evidence for the presence of sexual pheromones in free-spawning starfish. J. Exp. Mar. Biol. Ecol. 130:205–221.

Minale, L., Pizza, C., Riccio, R., and Zollo, F. 1982. Steroidal glycosides from starfishes. Pure Appl. Chem. 54:1935–1950.

Miyazawa, K., Higashiyama, M., Hori, K., Noguchi, T., Ito, K., and Hashimoto, K. 1987. Distribution of tetrodotoxin in various organs of the starfish *Astropecten polyacanthus*. Mar. Biol. 96:385–390.

Miyazawa, K., Noguchi, T., Maruyama, J., Jeon, J.K., Otsuka, M., and Hashimoto, K. 1985. Occurrence of tetrodotoxin in the starfishes *Astropecten polyacanthus* and *A. scoparius* in the Seto Inland Sea. Mar. Biol. 90:61–64.

Moore, R.E. 1978. Algal nonisoprenoids. *In* Marine natural products: chemical and biological perspectives, vol. 1, ed. P.J. Scheuer, pp. 44–124. New York: Academic Press.

Muehlstein, L.K., Porter, D., and Short, F.T. 1988. *Labyrinthula* sp., a marine slime mold producing the symptoms of wasting disease in eelgrass *Zostera marina*. Mar. Biol. 99:465–472.

Müller, D.G., Gassman, G., Boland, W., Marner, F.-J., and Jaenicke, L. 1981. *Dictyota dichotoma* (Phaeophyceae): identification of the sperm attractant. Science 212:1040–1041.

Müller, D.G., Gassman, G., and Luning, K. 1979. Isolation of a spermatozoid-releasing and -attracting substance from female gametophytes of *Laminaria digitata*. Nature 279:430–431.

Murata, M., Legrand, A.M., Ishibashi, Y., and Yasumoto, T. 1989. Structures of ciguatoxin and its congener. J. Am. Chem. Soc. 111:8929–8931.

Nakanishi, K. 1988. Characterization of factors that are intimately involved in the life of marine organisms. *In* Biomedical importance of marine organisms, ed. D. Fautin, pp. 59–67. Mem. Cal. Acad. Sci. Symp. No. 13. San Francisco: California Academy of Sciences.

Neidleman, S.L., and Geigert, J., eds. 1986. Biohalogenation: principles, basic roles, and applications. New York: Halsted Press.

Nitao, J.K., and Zangerl, A.R. 1987. Floral development and chemical defense allocation in wild parsnip (*Pastinaca sativa*). Ecology 68:521–529.

Okaichi, T., Anderson, D.M., and Nemoto, T., eds. 1989. Red tides: biology, environmental science, and toxicology. New York: Elsevier.

Okuda, R.K., Gulavita, N.K., Scheuer, P.J., Matsumoto, G.K., Rafii, S., and Clardy, J. 1986. Secondary metabolites of marine organisms. *In* New trends in natural products chemistry 1986, ed. A. Rahman, and P.W. LeQuesne, pp. 417–433. Amsterdam: Elsevier.

Pass, M.A., Capra, M.F., Carlisle, C.H., Lawn, I., and Coll, J.C. 1989. Stimulation of contractions in the polyps of the soft coral *Xenia elongata* by compounds extracted from other alcyonacean soft corals. Comp. Biochem. Physiol. C 94:677–681.

Paul, V.J., and Fenical, W. 1986. Chemical defense in tropical green algae, order Caulerpales. Mar. Ecol. Prog. Ser. 34:157–169.

Paul, V.J., and Fenical, W. 1987. Natural products chemistry and chemical defense in tropical marine algae of the phylum Chlorophyta. *In* Bioorganic marine chemistry, vol. 1, ed. P.J. Scheuer, pp. 1–29. Berlin: Springer-Verlag.

Peters, E.C. 1988. Symbiosis to pathology: are the roles of microorganisms as pathogens of coral reef organisms predictable from existing knowledge? *In* Proc. Sixth Int. Coral Reef Symp., vol. 1, pp. 205–210.

Peterson, C.H., and Renaud, P.E. 1989. Analysis of feeding preference experiments. Oecologia 80:82–86.

Ragan, M.A., and Glombitza, K. 1986. Phlorotannins, brown algal polyphenols. *In* Progress in phycological research, vol. 4, ed. F.E. Round, and D.J. Chapman, pp. 129–241. Bristol: Biopress.

Rhoades, D.F. 1979. Evolution of plant chemical defense against herbivores. *In* Herbivores: their interactions with secondary plant metabolites, ed. G.A. Rosenthal, and D.H. Janzen, pp. 3–54. New York: Academic Press.

Rhoades, D.F. 1985. Offensive-defensive interactions between herbivores and plants: their relevance in herbivore population dynamics and ecological theory. Am. Nat. 125:205–238.

Rhoades, D.F., and Cates, R.G. 1976. Toward a general theory of plant antiherbivore chemistry. Recent Adv. Phytochem. 10:168–213.

Robinson, T. 1979. The evolutionary ecology of alkaloids. *In* Herbivores: their interactions with secondary plant metabolites, ed. G.A. Rosenthal and D.H. Janzen, pp. 413–442. New York: Academic Press.

Rogers, S.D., and Paul, V.J. 1991. Chemical defenses of three *Glossodoris* nudibranchs and their dietary *Hyrtios* sponges. Mar. Ecol. Prog. Ser., in press.

Roughgarden, J., Gaines, S., and Possingham, H. 1988. Recruitment dynamics in complex life cycles. Science 241:1460–1466.

Russell, F.S. 1984. Marine toxins and venomous and poisonous marine plants and animals. Adv. Mar. Biol. 21:60–217.

Sammarco, P.W., and Coll, J.C. 1988. The chemical ecology of alcyonarian corals (Coelenterata: Octocorallia). *In* Bioorganic marine chemistry, vol. 2, ed. P.J. Scheuer, pp. 87–116. Berlin: Springer-Verlag.

Sammarco, P.W., and Coll, J.C. 1990. Lack of predictability in terpenoid function: multiple roles and integration with related adaptations in soft corals. J. Chem. Ecol. 16:273–289.

Scheuer, P.J., ed. 1978–1983. Marine natural products: chemical and biological perspectives, vols. 1–5. New York: Academic Press.

Scheuer, P.J., ed. 1987–1989. Bioorganic marine chemistry, vols. 1–3. Berlin: Springer-Verlag.

Scheuer, P.J. 1990. Some marine ecological phenomena: chemical basis and biomedical potential. Science 248:173–177.

Sheumack, D.D., Howden, M.E.H., Spence, I., and Quinn, R.J. 1978. Maculotoxin: a neurotoxin from the venom gland of the octopus *Hapalochlaena maculosa* identified as tetrodotoxin. Science 199:188–189.

Shimizu, Y. 1984. Paralytic shellfish poisons. Prog. Chem. Org. Nat. Prod. 45:235–264.

Southwood, T.R.E., Brown, V.K., and Reader, P.M. 1986. Leaf palatability, life expectancy and herbivore damage. Oecologia 70:544–548.

Spencer, K.C., ed. 1988. Chemical mediation of coevolution. San Diego: Academic Press.

Starr, M., Himmelman, J.H., and Therriault, J.C. 1990. Direct coupling of marine invertebrate spawning with phytoplankton blooms. Science 247:1071–1074.

Steidinger, K.A., and Baden, D.G. 1984. Toxic marine dinoflagellates. *In* Dinoflagellates, ed. D. Spector, pp. 201–261. Orlando: Academic Press.

Stonik, V.A., and Elyakov, G.B. 1988. Secondary metabolites from echinoderms as chemotaxonomic markers. *In* Bioorganic marine chemistry, vol. 2, ed. P.J. Scheuer, pp. 43–86. Berlin: Springer-Verlag.

Swain, T. 1979. Tannins and lignins. *In* Herbivores: their interactions with secondary plant metabolites, ed. G.A. Rosenthal, and D.H. Janzen, pp. 657–682, New York: Academic Press.

Tachibana, K. 1988. Chemical defenses in fishes. *In* Bioorganic marine chemistry, vol. 2, ed. P.J. Scheuer, pp. 117–130. Heidelberg: Springer-Verlag.

Thuesen, E.V., and Kogure, K. 1989. Bacterial production of tetrodotoxin in four species of Chaetognatha. Biol. Bull. 176:191–194.

Thuesen, E.V., Kogure, K., Hashimoto, K., and Nemoto, T. 1988. Poison arrowworms: a tetrodotoxin venom in the marine phylum Chaetognatha. J. Exp. Mar. Biol. Ecol. 116:249–256.

Van Alstyne, K.L. 1986. Effects of phytoplankton taste and smell on feeding behavior of the copepod *Centropages hamatus*. Mar. Ecol. Prog. Ser. 34:187–190.

Ward, J.E., and Targett, N.M. 1989. Influence of marine microalgal metabolites on the feeding behavior of the blue mussel *Mytilus edulis*. Mar. Biol. 101:313–321.

Wilson, E.O. 1970. Chemical communication within animal species. *In* Chemical ecology, ed. E. Sondheimer and J.B. Simeone, pp. 133–155. New York: Academic Press.

Yasumoto, T., Murata, M., Oshima, Y., Sano, M., Matsumoto, G.K., and Clardy, J. 1985. Diarrhetic shellfish toxins. Tetrahedron 41:1019–1026.

Yasumoto, T., Yasumura, D., Yotsu, M., Michishita, T., Endo, A., and Kotaki, Y. 1986. Bacterial production of tetrodotoxin and anhydrotetrodotoxin. Agric. Biol. Chem. 50:793–795.

Chapter 1
Seaweed Chemical Defenses on Coral Reefs

VALERIE J. PAUL

In contrast to the abundance of seaweeds found in many temperate marine habitats, seaweeds are often relatively inconspicuous on
coral reefs. Intense grazing by fishes, sea urchins, and other herbivores
is largely responsible for the low biomass of marine plants found in
many reef habitats (Bakus 1964; Borowitzka 1981; Steneck 1988). Most
tropical seaweeds are common only in habitats where herbivorous
fishes and urchins are less abundant (Hay 1984b). Coral reef seaweeds
that survive the grazing activities of large herbivores are often chemically or morphologically defended (Steneck 1986, 1988; Hay and Fenical 1988). In this chapter, I discuss the chemical defenses tropical seaweeds use against herbivores, focusing on effects toward large, mobile
grazers such as fishes and sea urchins.

Ecology of Herbivory on Coral Reefs

Herbivory on some coral reefs is probably more intense than in any
other terrestrial or marine habitat (Hatcher and Larkum 1983; Lewis
1986; Carpenter 1986; Hay 1991a). Carpenter (1986) showed that an annual average of 97% of the algal turf production in the back-reef/reef-
crest habitat was removed by herbivores on St. Croix. Many other
studies have shown that large herbivores, particularly grazing fishes,
directly affect the species composition and abundance of benthic marine algae on tropical reefs and seagrass beds (Stephenson and Searles
1960; Randall 1961, 1965, 1974; Earle 1972; John and Pople 1973; Vine

24

1974; Mathieson et al. 1975; Ogden 1976; Wanders 1977; Ogden and Lobel 1978; Brock 1979; Hay 1981a; Hixon and Brostoff 1983; Carpenter 1986; Lewis 1986; Morrison 1988; Coen and Tanner 1989). Herbivory is therefore an important pressure selecting for the evolution of defensive characteristics in tropical seaweeds.

A variety of herbivore taxa are ecologically important in the tropics. Approximately 25% of the number and biomass of fish species on tropical reefs consists of herbivorous fishes of the families Scaridae (parrotfish), Acanthuridae (surgeonfish), Pomacentridae (damselfish), Kyphosidae (chubs), and, in the Pacific, the Siganidae (rabbitfish). These herbivores, as well as some other groups of omnivorous fishes, derive their energy primarily from benthic plants (Bakus 1964, 1966, 1969; Hiatt and Strasburg 1960; Randall 1967; Hobson 1974; reviewed by Horn 1989).

Grazing by sea urchins also significantly affects the distribution and abundance of marine macroalgae in some tropical habitats, especially in the Caribbean, as demonstrated by removal and caging experiments (Dart 1972; Ogden et al. 1973; Sammarco et al. 1974; Lawrence 1975; Sammarco 1980, 1982; Carpenter 1981, 1986; Birkeland 1989). (The proliferation of sea urchins on some Caribbean reefs may be the result of human activities such as overfishing [Hay 1984a], which reduces such natural predators as triggerfishes and wrasses.) In February 1983, a massive die-off of the Caribbean urchin *Diadema antillarum* occurred (Bak et al. 1984; Lessios et al. 1984; Lessios 1988). Changes in the abundance, composition, and productivity of reef seaweeds followed, indicating that sea urchins have an important effect on algal community structure (Carpenter 1985, 1988; Hughes et al. 1987; Lessios 1988; Levitan 1988; Morrison 1988). Sea urchins may compete with herbivorous fishes for food in habitats where urchins are abundant (Hay and Taylor 1985).

Macroalgal biomass is low in many reef habitats as a result of this intense herbivory (Borowitzka 1981; Huston 1985; Hackney et al. 1989). Herbivorous fishes and urchins demonstrate distinct feeding preferences in field and laboratory assays (Earle 1972; Tsuda and Bryan 1973; Mathieson et al. 1975; Ogden 1976; Ogden and Lobel 1978; Lobel and Ogden 1981; Littler et al. 1983; Hay 1984b; Lewis 1985; Paul and Hay 1986; Morrison 1988). Stomach content analyses also indicate that certain algae are preferred and some abundant species are avoided under natural foraging conditions (Hiatt and Strasburg 1960; Randall 1967; Hobson 1974; Lawrence 1975).

Seaweeds have several mechanisms for resisting herbivory (Lubchenco and Gaines 1981; Duffy and Hay 1990; Hay 1991b). Recent studies have demonstrated that many seaweeds can deter herbivores by using chemical and morphological defenses (Steneck 1986; Hay and Fenical 1988) or by associating with deterrent seaweeds or other benthic organisms that reduce foraging (Hay 1985, 1986; Littler et al. 1986, 1987; Pfister and Hay 1988). Morphological defenses such as calcification and toughness have been discussed previously (Littler and Littler 1980; Steneck and Watling 1982; Littler et al. 1983; Steneck 1986, 1988). Chemical defenses of seaweeds have been reviewed recently (Hay and Fenical 1988; Van Alstyne and Paul 1988) and will be discussed more thoroughly in this chapter. Often, several herbivore resistance mechanisms may function simultaneously (Littler and Littler 1980; Hay 1984b; Paul and Hay 1986), and the importance of multiple defenses may be very significant in herbivore-rich tropical waters. Several common tropical seaweeds combine chemical and morphological defenses (Paul and Fenical 1983, 1986; Hay 1984b; Paul and Hay 1986; Hay et al. 1988a; Paul and Van Alstyne 1988a, 1988c).

Generally, grazing by fishes is most intense on reef slopes, and only seaweeds that grow rapidly to replace tissues lost to herbivores or are chemically or morphologically defended occur there (Hay 1981a, 1981b, 1984b, 1985; Hay et al. 1983; Lewis 1985, 1986; Lewis and Wainwright 1985; Coen and Tanner 1989). Rapidly growing filamentous turf algae and calcified crustose algae predominate in these habitats. Reef flats, reef crests, seagrass beds, lagoons, mangroves, and deep-reef sand plains are habitats where herbivorous fish densities are relatively low (Hay 1981a, 1981b, 1984b; Hay et al. 1983; Russ 1984a, 1984b; Taylor et al. 1986), and many seaweeds avoid herbivory by living in these areas. Seaweeds that normally grow in these habitats can usually grow on reef slopes if protected from grazers, but they are consumed rapidly if left unprotected (Hay 1981a, 1981b, 1984b; Hay et al. 1983; Lewis 1985, 1986).

The effects of less conspicuous species of tropical herbivores such as amphipods, crabs, polychaetes, and molluscs (termed "mesograzers" by Hay et al. 1987a) on seaweeds are less well known (Brawley and Adey 1981a, 1981b; Carpenter 1986). Hay et al. (1987a) hypothesized that mesograzers should be more likely to consume seaweeds that are resistant to macroherbivores because mesograzers are subject to heavy predation by grazing fishes. Mesograzers are especially likely to be eaten if they live on preferred seaweeds. Evidence that amphipods, ascoglossans, and polychaetes prefer to live on and eat low-preference, chemically defended seaweeds has been accumulating (Hay et al. 1987a, 1988b, 1988c, 1989, 1990a, 1990b; Paul and Van Alstyne 1988b; Hay, this volume).

Chemical Defenses of Tropical Seaweeds

More than 600 natural products have been isolated from seaweeds, and the majority of these have come from tropical algae (Faulkner 1984, 1986, 1987, 1988a). In general, these compounds occur in relatively low concentrations, ranging from 0.2% to 2% by dry weight (Paul and Fenical 1986; Hay and Fenical 1988). Except for metabolites from phytoplankton and blue-green algae (cyanobacteria), very few nitrogenous compounds have been isolated from the algae. The majority of macroalgal compounds are terpenoids, especially sesqui- and diterpenoids. Acetogenins (acetate-derived metabolites), including unusual fatty acids, constitute another common class of seaweed secondary metabolites. Most of the remaining metabolites result from mixed biosynthesis and are often composed of terpenoid and aromatic portions. The greatest variety of secondary metabolites is probably found among the red algae (Rhodophyta), in which all classes of compounds are represented and many metabolites are halogenated (Fenical 1975; Faulkner 1984, 1986, 1987, 1988a).

Many possible defensive functions for algal secondary metabolites have been proposed, including antimicrobial, antifouling, and antiherbivore. To date, the role of these compounds as defenses against herbivores is best known. Recent studies have clearly shown that many seaweed natural products function as feeding deterrents (reviewed by Hay and Fenical 1988); however, many compounds may also have other roles or may function simultaneously as defenses against pathogens, fouling organisms, and herbivores, thereby increasing the adaptive value of these metabolites. Some algal secondary metabolites do show antimicrobial (Almodovar 1964; Caccamese et al. 1980; Hodgson 1984) or antifouling (Sieburth and Conover 1965; McLachlan and Craigie 1966; Al-Ogily and Knight-Jones 1977) effects.

A consideration of multiple functions is important because even though algal secondary metabolites may function as defenses against herbivores, they may have evolved for other reasons, such as resistance to pathogens or competitors. Many tropical algae are evolutionarily older than the herbivorous fishes that graze on them. For example, ancestral perciform fishes are known from the upper Cretaceous (70–80 million years ago; Lagler et al. 1977), whereas fossils of calcareous green algae are known from the Cambrian (500 million years ago; Bold and Wynne 1978).

The common method of testing for feeding deterrent effects has been to incorporate seaweed extracts or isolated metabolites at natural concentrations into a palatable diet—either a preferred seaweed or an artificial diet—and then to compare feeding rates of the grazers on treated foods with those on appropriate controls (McConnell et al. 1982; Targett et al. 1986; Hay et al. 1987b; Paul 1987). Deterrent effects observed in these assays appear to be based primarily on the taste of the treated food. If a compound deters an herbivore, the degree of avoidance is usually related to the concentration of the extract or metabolite in the diet. Higher concentrations result in more pronounced deterrent effects (Hay and Fenical 1988; Paul et al. 1988). These methods do not assess toxicity or other physiological effects on the consumers or possible detoxification methods by herbivores.

Usually the presence or absence of deterrent secondary metabolites in seaweeds correlates well with seaweeds' susceptibility to herbivores. Seaweeds that are least susceptible to grazing fishes employ chemical or morphological defenses or both (Hay 1984b; Paul and Hay 1986; Paul and Potter 1992), although, some herbivorous fishes do consume chemically rich seaweeds. For instance, the rabbitfish *Siganus argenteus* readily consumes the green seaweed *Chlorodesmis fastigiata* (Tsuda and Bryan 1973; Paul et al. 1990), which is avoided by another rabbitfish, *S. spinus*, and many other herbivorous fishes on Guam (Paul and Potter 1992). It is not known whether *S. argenteus* detoxifies or simply tolerates the diterpenoid metabolites produced by *Chlorodesmis*. Another green seaweed, *Caulerpa racemosa*, is readily consumed by many herbivorous fishes (Paul and Hay 1986; Paul and Potter 1992), although the alga produces sesquiterpenoid metabolites in relatively high concentrations (1–2% of dry weight; Meyer and Paul, submitted). Neither the extract of *Caulerpa* nor the major metabolite caulerpenyne (1)

1

deters many of Guam's herbivorous fishes (Paul 1987; Paul et al. 1988, 1990; Wylie and Paul 1988; Meyer and Paul, submitted).

It is not clear what chemical features determine the degree of deterrency or toxicity of any particular secondary metabolite. Toxicity and deterrency are not intrinsic properties of any compound; they result from physiological interactions between a metabolite and its

Table 1.1

Effects of algal secondary metabolites on different coral reef herbivores

	Caribbean herbivores				Pacific herbivores			
Compound	Sparisoma radians (fish)[a]	Field assays (fish)	Diadema (urchin)[b]	Amphipods	Field (Guam)	Field (Great Barrier Reef)[c]	Zeb-rasoma (fish)[d]	Siganus argenteus (fish)[e]
Chlorophytes caulerpenyne (1)	+	−[f]			−[f]		−	−
halimedatetraacetate (10)	+	−[f]	+[h]		+[g]			
halimedatrial (9)					+[g]			

caulerpenyne (1)

halimedatetraacetate (10)

halimedatrial (9)

Table 1.1—*Continued*

	Caribbean herbivores				Pacific herbivores			
Compound	*Sparisoma radians* (fish)[a]	Field assays (fish)	*Diadema* (urchin)[b]	Amphipods	Field (Guam)	Field (Great Barrier Reef)[c]	*Zeb-rasoma* (fish)[d]	*Siganus argenteus* (fish)[e]
udoteal (2)		–[f]			+/–[i]		+	–
chlorodesmin (4)					+[i]	–	+	+
cymopol (12)		+[b]				–		

udoteal (2)

chlorodesmin (4)

A

cymopol (12)

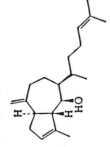

Phaeophytes
pachydictyol-A (8)

stypotriol (13)

dictyopterene A & B (14)

+^b + −^j + + −

+^b + + − +

+^k −^k +

31

Table 1.1—*Continued*

Compound	Caribbean herbivores				Pacific herbivores			
	Sparisoma radians (fish)[a]	Field assays (fish)	*Diadema* (urchin)[b]	Amphipods	Field (Guam)	Field (Great Barrier Reef)[c]	*Zebrasoma* (fish)[d]	*Siganus argenteus* (fish)[e]
Rhodophytes isolaurinterol (15)	+[b]		+					
aplysin (16)	−[b]		−					
elatol (5)	+[b]		+		+[j]	+	−	−

ochtodene (17)

chondrocole C (18)

palisadin-A (6)

Note: Experiments with individual species were conducted in aquaria. Field assays were done in reef habitats with compounds tested on natural populations of herbivorous fishes. + = feeding deterrent; − = not a significant feeding deterrent; A = significant attractant.

Sources: [a]Targett et al. 1986; [b]Hay et al. 1987b; [c]Hay et al. 1988d; [d]Wylie and Paul 1988; [e]Paul et al. 1990; [f]Paul, pers. observ.; [g]Paul and Van Alstyne 1988a; [h]Hay et al. 1988a; [i]Paul 1987; [j]Paul et al. 1988; [k]Hay et al. 1988c; [l]Paul et al. 1987a.

33

consumer. The same metabolite may show pronounced differences in its effects even on closely related species of herbivores. Some metabolites inhibit feeding by most herbivores, whereas other compounds deter only a few herbivores (Table 1.1). Compounds that differ only slightly in chemical structure can vary greatly in their deterrent effects (Hay et al. 1987b, 1988d; Paul et al. 1987a, 1988, 1990). For example, the closely related green algal metabolites udoteal (2), flexilin (3), chlorodesmin (4),

2

3

4

and caulerpenyne (1) differ in their feeding deterrent effects even though all contain the same major bis-enol acetate functional group (Paul 1987; Paul et al. 1988; Table 1.1). Similarly, halogenation in red algal metabolites does not necessarily enhance their feeding deterrent effects. Of three halogenated sesquiterpenoids from *Laurencia*, elatol (5) is broadly effective as a feeding deterrent, palisadin-A (6) deters some herbivorous fishes but not others, and aplysistatin (7) does not appear to have feeding deterrent effects (Table 1.1; Hay et al. 1988d; Paul et al. 1988).

5   6   7

Thus, predictions about the toxic or deterrent effects of particular secondary metabolites may be difficult to make based on chemical structures or results of pharmacological assays. Field and laboratory assays with natural predators are necessary to examine these ecological interactions.

Little is known about the physiology of marine herbivores with regard to the way secondary metabolites are detected or metabolized. It is likely that many marine herbivores detect algal secondary metabolites by taste rather than smell, because many compounds are sequestered in membrane-bound vesicles in cells and are not exuded into seawater (Pederson et al. 1980; Young et al. 1980). Herbivores would therefore have to bite the plant to perceive deterrent compounds.

Steinberg (1985) suggested that because temperate brown algal polyphenolics consistently function as feeding deterrents against marine invertebrate herbivores in North America, they should also affect the fitness of the herbivores by some toxic, growth-inhibiting, or other deleterious effects if the compounds are to be effective over evolutionary time. This argument should hold for feeding deterrent effects of any seaweeds. Thus, feeding preferences of herbivores should be related to their physiological tolerances for the chemical defenses of different seaweeds and their growth and performance on different algal diets. Hay et al. (1987a) showed that when the brown algal metabolite pachydictyol-A (8)

8

was fed to temperate omnivorous pinfish at 1.0% of food dry mass, the fish's growth rate was reduced by almost 50% relative to controls. The fish appeared to recognize the taste of pachydictyol-A after a few days, avoided it, and had to be tricked into eating it by dipping food pellets into a slurry of tuna-flavored cat food.

Several studies have shown that chemically rich tropical green algae are poor diets for some tropical herbivores. The parrotfish *Sparisoma radians* had very low survivorship (equivalent to starvation) when fed diets of *Caulerpa mexicana, Halimeda incrassata,* and *Penicillus pyriformis* (Lobel and Ogden 1981). Lobel and Ogden (1981) attributed this mortality to toxins in *Caulerpa* and calcification in *Halimeda* and *Penicillus*; however, it is likely that the toxic terpenoids present in all three species were responsible for the high mortality. Targett et al. (1986) later demonstrated that the major terpenoid metabolites in *Caulerpa* and *Halimeda* deterred feeding by *S. radians. Caulerpa racemosa* was also barely susceptibile to natural populations of parrotfishes on Jamaican reefs (Morrison 1988). Balazs (1982) showed that juvenile green turtles in the northwestern Hawaiian Islands fed on *C. racemosa* and other unpreferred algae only when other more desirable algae were absent. This population of turtles had lower growth rates than turtles from the southern Hawaiian Islands that fed on preferred algae.

Metabolites that are toxic or deterrent to generalist herbivores such as fishes and sea urchins are ineffective against some specialist herbivores and may even be feeding attractants. Specialist herbivorous molluscs such as sea hares and ascoglossans selectively consume chemically rich seaweeds and often concentrate the seaweed secondary metabolites within their own bodies (reviewed by Hay and Fenical 1988; Faulkner, this volume). These metabolites are exuded when the molluscs are attacked, and they appear to deter predators (Faulkner 1984, 1988b; Hay and Fenical 1988; Paul and Van Alstyne 1988b; Cimino et al. 1990; Hay et al. 1990b; Paul and Pennings 1991). Some ascoglossans appear to biosynthesize their own metabolites (Ireland and Faulkner 1981; Dawe and Wright 1986; Roussis et al. 1990). Ascoglossans feed suctorially and can retain photosynthetically functional chloroplasts from their host algae (Trench 1975, 1980). Thus, some ascoglossans may be capable of synthesizing their chemical defenses from primary metabolites derived from photosynthesis (Ireland and Faulkner 1981). Other small generalized herbivores (mesograzers) such as amphipods, crabs, and polychaetes may also preferentially consume seaweeds that are chemically defended from fishes (Hay et al. 1987a, 1988b, 1988c; Hay and Fenical 1988; Hay 1991a, this volume).

Amphipods, crabs, and polychaetes do not appear to sequester algal secondary metabolites.

Variation in Secondary Metabolite Production

Given the variability of herbivores' responses to seaweed chemical defenses, it is not surprising that tropical seaweeds produce a variety of secondary metabolites. Different herbivores have different methods of foraging and different digestive physiologies; thus, the effectiveness of any single plant defense should decrease as the diversity of herbivore types within a community increases (Lubchenco and Gaines 1981; Gaines 1985). It is possible that seaweeds produce an assortment of secondary metabolites because different compounds deter different herbivores. This may be especially true for tropical seaweeds exposed to a diverse assemblage of herbivores. Metabolite types and concentrations vary within and among plants in a population, with the age of plants, and among conspecific populations of seaweeds (reviewed below, and in Hay and Fenical 1988; Van Alstyne and Paul 1988).

Variation in secondary metabolites occurs at a number of levels. Differences in concentrations and types of chemical defenses within plants occur in *Halimeda*, in which more toxic and deterrent compounds are allocated to areas of new growth (Hay et al. 1988a; Paul and Van Alstyne 1988a). Chemical variation among individuals also occurs in *Halimeda* spp. (Paul 1985; Paul and Van Alstyne 1988a, 1988c) and the brown alga *Stypopodium zonale* (Gerwick et al. 1985). Some of this variation may be related to the age of plants (Hay et al 1988a; Paul and Van Alstyne 1988a). Variations in types and concentrations of secondary metabolites also occur among populations of seaweeds growing in different reef habitats. Populations of *Halimeda* from habitats in which herbivory is intense tend to contain higher levels of the more potent deterrent halimedatrial (9)

9

than do populations from areas of low herbivory (Paul and Van Alstyne 1988a). Other green algae such as *Penicillus, Udotea, Rhipocephalus,* and *Caulerpa* also often produce higher concentrations or different types of secondary metabolites in populations from herbivore-rich reef habitats than in populations from herbivore-poor areas such as reef flats or seagrass beds (Paul and Fenical 1986, 1987; Paul et al. 1987b; pers. observ.). We do not know if these variations in concentrations and types of secondary metabolites result from herbivore-induced chemical defenses, localized selection resulting in high levels of defense, or other factors not related to herbivory.

If seaweeds respond to damage by increasing or changing secondary metabolites (inducible defenses), then this could generate the patterns of variation described above. Herbivore-induced chemical defenses have been demonstrated only for one temperate alga, *Fucus distichus* (Van Alstyne 1988), in which levels of polyphenolic compounds increased in response to mechanical damage. Similar experiments with green algae of the genera *Halimeda*, *Udotea*, and *Caulerpa* did not result in any change in levels of chemical defenses in response to mechanical injury for periods ranging from several hours to several weeks (Paul and Van Alstyne, submitted). These species appeared to produce consistently high levels of secondary metabolites (constitutive defenses).

Although the tropical green seaweeds do not increase their production of secondary metabolites in response to damage, several species of *Halimeda* and *Udotea* can rapidly convert a less toxic and deterrent terpenoid metabolite to a more deterrent compound. This process, which has been termed "activation" (Paul and Van Alstyne 1988c), occurs within seconds of tissue injury and appears to be enzymatically mediated. Only the portion of the plant in the immediate vicinity of the injury is affected. *Halimeda* species convert the less deterrent major metabolite halimedatetraacetate (10) to the more toxic and deterrent halimedatrial (9) (Paul and Van Alstyne 1988c, submitted).

Udotea flabellum converts the less deterrent metabolite udoteal (2) to the more toxic and deterrent petiodial (11) (pers. observ.).

These conversions occur after any mechanical injury, and could occur when a fish bites or chews on the plants. Activation is also common in some terrestrial plants; examples include plants that produce HCN from organic precursors (Conn 1979) and plants that convert glucosinolates to thiocyanates and isothiocyanates (Chew 1988) after tissue damage. In these cases, precursor compounds are compartmentalized and physically separated from the enzymes that activate them.

Because the process of activation is so rapid, it may be a common defense mechanism in habitats where herbivory is intense and herbivores are large and mobile; for example, on reef slopes. In these habitats, herbivory is intense but highly variable over short periods (i.e., when grazing fishes bite plants and swim away), and herbivores are large and can rapidly and completely consume plants. In contrast, herbivore-induced defenses may occur in tropical habitats such as reef flats, where grazing intensity is variable over longer periods (i.e., seasonal variation or recruitment events), predators are small relative to prey size (e.g., amphipods, molluscs), and partial predation is common. These predictions for induced and activated defenses still need to be tested for seaweeds in different coral reef habitats.

Variations in secondary metabolite production also occur over geographic scales. Most seaweed secondary metabolites appear to be produced by tropical or subtropical species (Faulkner 1984, 1986, 1987, 1988a). This appears to be related to the high levels of herbivory found on many tropical reefs. On the other hand, tropical brown seaweeds of the order Fucales do not produce the high levels of polyphenolic metabolites that related temperate species do. Fucales and other brown seaweeds in tropical habitats appear to contain low levels of polyphenolics (Steinberg 1986, 1989, this volume; Steinberg and Paul 1990; Van Alstyne and Paul 1990). The reasons for this geographical variation in polyphenolics among the brown seaweeds is not understood. Many tropical brown seaweeds, especially those in the order Dictyotales, produce secondary metabolites that are terpenoids or products of mixed biosynthesis. It is possible that physiological constraints or nutrient limitation may restrict the production of polyphenolics by tropical brown seaweeds.

Effects of Chemical Defense on Coral Reef Community Structure

Many chemically rich seaweeds probably persist in reef habitats because of the toxic or deterrent secondary metabolites they produce or because of their combined chemical and morphological defenses. The defensive abilities of seaweeds, however, can vary both spatially and temporally on coral reefs. Most algal secondary metabolites are effective chemical defenses against certain herbivores but not others. Similarly, morphological defenses such as calcification or toughness may be effective against some herbivorous fishes, but others, such as the parrotfish, can readily consume calcified prey. Therefore, tropical seaweeds that are effectively defended in some reef habitats or at some times of the year may be consumed in others because of differences in herbivore community structure. This is certainly true if we compare seaweed communities in reef habitats where herbivory is high with the higher diversity and biomass of seaweeds found in lagoons and reef flats, where herbivory is low (Hay 1984b, 1985; Russ 1984a, 1984b; Lewis 1985, 1986; Steneck 1988). But differences in seaweed communities may also occur among high herbivory habitats when herbivorous fishes vary in numbers and species composition (Russ 1984a, 1984b; Choat and Bellwood 1985; Lewis and Wainwright 1985; Scott and Russ 1987).

I suggest that differences in herbivore population biology and community structure among reef habitats, and differences in interactions between herbivores and secondary metabolites, can lead to variation in seaweed community structure. Mechanisms for examining these differential effects include testing seaweed susceptibility (e.g., Hay 1984b; Lewis 1985; Paul and Hay 1986) and feeding deterrent effects of seaweed secondary metabolites in a variety of reef habitats. Deterrence assays conducted in the field test the effectiveness of an extract or secondary metabolite against natural populations of herbivorous fishes within a reef community. Results of field assays with extracts and isolated metabolites in different reef habitats should indicate whether the effectiveness of seaweed chemical defenses varies among habitats. For example, udoteal (2) deterred herbivorous fishes in field assays on some reefs on Guam but not others (Paul 1987).

Differences among seaweed communities may be observed between reefs in the Caribbean and Pacific. For example, some species of *Laurencia* and *Dictyota* are common on Caribbean reefs in habitats where herbivory is high, and they are relatively resistant to grazing (Ogden 1976; Littler et al. 1983; Hay 1984b; Paul and Hay 1986). In contrast, on Guam and in other parts of Micronesia, *Dictyota* and *Laurencia* are rarely found in reef-slope habitats, but instead are restricted to reef flats and lagoons (Tsuda and Wray 1977; Paul 1987, pers. observ.). Similar secondary metabolites are produced by many Caribbean and Pacific species of *Dictyota* and *Laurencia*, and regional differences in species distribution suggest that the chemical defenses of these seaweeds against large herbivores may not be as effective on reefs in Guam and Micronesia as they are in the Caribbean. These differences in effectiveness may be a result of the greater diversity of herbivorous fishes in the western Pacific; siganids (rabbitfish) and several genera of surgeonfish (*Naso*, *Zebrasoma*) are important consumers in the Indo-Pacific but are not found in the Caribbean. For example, elatol (5) and palisadin-A (6) (from *Laurencia* spp.) did not deter the rabbitfish *Siganus argenteus* or the surgeonfish *Zebrasoma flavescens* (Table 1.1; Paul et al. 1988), and pachydictyol-A (8) (from *Dictyota* spp.) did not deter *Siganus argenteus* or the surgeonfish *Naso lituratus* on Guam (Table 1.1; Paul et al. 1988, 1990; pers. observ.). In Caribbean field assays, however, both elatol and pachydictyol-A were deterrent against natural populations of herbivorous fishes and the sea urchin *Diadema antillarum* (Table 1.1; Hay et al. 1987b). As the diversity of herbivore species increases, the probability of having herbivores that are not affected by any particular chemical defense undoubtedly increases as well.

Herbivore numbers and community structure may also vary with seasonal or episodic recruitment events (Sale 1980, 1988; Sale and Douglas 1984). For example, Carpenter (1986) showed that herbivorous fish abundance varied significantly over time on a reef in St. Croix. Peak abundances occurred in September, when juvenile parrotfish represented 95% of the herbivorous fish species present. Mesograzer abundances and the relative composition of herbivorous amphipods, gastropods, and polychaetes also varied greatly over time in back-reef and lagoon habitats (Carpenter 1986). On Guam, two species of rabbitfish (*Siganus spinus* and *S. argenteus*) recruit on reef flats each year in April and May (Tsuda and Bryan 1973; Kami and Ikehara 1976). The magnitude of the recruitment varies annually. In some years almost no juveniles are observed, but in other years up to 13 million juveniles have been estimated on reef flats on Guam during a one-week run (Tsuda and Bryan 1973). When the juveniles recruit in large numbers, they eat almost all of the seaweeds from reef-flat habitats, including normally unpalatable species such as *Halimeda* and *Avrainvillea* (pers. observ.). Habitats that normally experience low levels of herbivory can suddenly be overwhelmed by juvenile rabbitfish, and most species of seaweed disappear. Thus, the effects of massive juvenile recruitment could provide the most striking demonstration of how herbivore populations control the structure of seaweed communities. Only seaweeds that are effectively defended chemically or morphologically would be expected to persist when juveniles are very numerous.

Conclusions

Recent studies have clearly demonstrated that many secondary metabolites produced by tropical seaweeds function as chemical defenses against herbivores in reef habitats where grazing is intense. Not all compounds are equally deterrent. Some appear to be broadly deterrent against a variety of herbivores, while others may deter only a few species or none at all. Compounds that vary only slightly in their chemical structures may have very different toxic or deterrent effects. The same metabolite may differ in its effects on even related species of herbivores. Additionally, compounds that are deterrent or toxic to generalists may be ineffective at deterring specialists or may even be feeding attractants.

Given this variability in the responses of herbivores to seaweed chemical defenses, it is not surprising that a great deal of chemical variation occurs in the secondary metabolites produced by seaweeds. Metabolite types and concentrations vary within and among plants in a population, according to the age of plants, and among conspecific populations of seaweeds. Variation in the types and concentrations of secondary metabolites may result from increased production of compounds due to increased grazing over time (induction), from rapid conversion of compounds from less deterrent to more deterrent metabolites when plants are injured (activation), or from localized selection by herbivores.

Clearly, chemically mediated interactions between herbivores and marine plants are complex and depend on the responses of different species of herbivores and the types and concentrations of secondary metabolites produced by different populations of seaweeds. We now know that chemical variation occurs widely in tropical seaweeds, but we do not know how important biological and physical environmental causes of this variation are for different seaweeds. Variability in seaweed chemistry could result from previous grazing history, from physical environmental factors such as light and nutrients, from interactions with competitors or pathogens, from genetic differences among algal populations, or as a result of localized selection. Chemically rich seaweeds may survive grazing by herbivorous fishes in some reef habitats and not others because of differences among fish communities.

The physiology of the seaweeds producing chemical defenses and the herbivores consuming them is not well known. Biosynthesis, storage, and turnover of secondary metabolites might vary according to whether seaweeds produce constitutive, inducible, or activated defenses. The turnover rates of different secondary metabolites are important considerations in estimating the cost of a particular defense, but these have not been measured in seaweeds. Almost nothing is known about the physiological mechanisms of secondary metabolites' toxicity. Pharmacological assays have shown that some marine secondary metabolites function as neurotoxins or cardiac toxins in laboratory animals, but mechanisms of toxicity may differ when a metabolite is ingested. Some generalist herbivores can consume chemically rich seaweeds, but whether the secondary metabolites are absorbed, detoxified, or tolerated is not known.

Considerable progress has been made over the past few years in the area of marine chemical ecology of plant-herbivore interactions, but many questions remain regarding the production, modes of action, and ecology of seaweed secondary metabolites. Research in this area is interdisciplinary in scope, and continued collaboration between chemists and biologists will be important for understanding the chemicals involved in these plant-herbivore interactions and the biology of coral reefs, seaweeds, and herbivores.

Acknowledgments My studies of marine plant-herbivore interactions on Guam were funded by the National Science Foundation (OCE-8600998) and the National Oceanic and Atmospheric Administration, Office of Sea Grant (Institutional Grant no. NA85AA-D-SG082). I am grateful to Steve Nelson, Peter Steinberg, and Mark Hay for helpful comments on drafts of this chapter.

References

Almodovar, L.R. 1964. Ecological aspects of some antibiotic algae in Puerto Rico. Bot. Mar. 6:143–146.
Al-Ogily, S.M., and Knight-Jones, E.W. 1977. Anti-fouling role of antibiotics produced by marine algae and bryozoans. Nature 265:728–729.
Bak, R.P.M., Carpay, M.J.E., and de Ruyter van Stevenick, E.D. 1984. Densities of the

sea urchin *Diadema antillarum* before and after mass mortalities on the coral reefs of Curaçao. Mar. Ecol. Prog. Ser. 17:105–108.

Bakus, G. 1964. The effects of fish grazing on invertebrate evolution in shallow tropical waters. Allan Hancock Found. Occas. Pap. 27:1–29.

Bakus, G. 1966. Some relationships of fishes to benthic organisms on coral reefs. Nature 210:280–284.

Bakus, G. 1969. Energetics and feeding in shallow marine waters. Int. Rev. Gen. Exp. Zool. 4:275–369.

Balazs, G.H. 1982. Growth rates of immature green turtles in the Hawaiian archipelago. *In* Biology and conservation of sea turtles (Proc. of World Conference on Sea Turtle Conservation, Nov. 1979, Washington D.C.), ed. K.A. Bjorndal, pp. 117–125. Washington, D.C.: Smithsonian Institution Press.

Birkeland, C. 1989. The influence of echinoderms on coral-reef communities. *In* Echinoderm studies, vol. 3, ed. M. Jangoux and J.M. Lawrence, pp. 1–79. Rotterdam: A.A. Balkema.

Bold, H.C., and Wynne, M.J. 1978. Introduction to the algae. Englewood Cliffs, N.J.: Prentice-Hall.

Borowitzka, M.A. 1981. Algae and grazing in coral reef ecosystems. Endeavour 5:99–106.

Brawley, S.H., and Adey, W.H. 1981a. The effect of micrograzers on algal community structure in a coral reef microcosm. Mar. Biol. 61:167–177.

Brawley, S.H., and Adey, W.H. 1981b. Micrograzers may affect macroalgal density. Nature 292:177.

Brock, R.E. 1979. An experimental study on the effects of grazing by parrotfishes and role of refuges in benthic community structure. Mar. Biol. 51:381–388.

Caccamese, S., Azzolina, R., Furnari, G., Cormaci, M., and Grasso, S. 1980. Antimicrobial and antiviral activities of extracts from Mediterranean algae. Bot. Mar. 23:285–288.

Carpenter, R.C. 1981. Grazing by *Diadema antillarum* Phillipi and its effect on the benthic algal community. J. Mar. Res. 39:749–765.

Carpenter, R.C. 1985. Sea urchin mass mortality: effects on reef algal abundance, species composition, and metabolism of other coral reef herbivores. *In* Proc. Fifth Int. Coral Reef Symp., vol. 4, pp. 53–60.

Carpenter, R.C. 1986. Partitioning herbivory and its effects on coral reef algal communities. Ecol. Monogr. 56:343–363.

Carpenter, R.C. 1988. Mass mortality of a Caribbean sea urchin: immediate effects on community metabolism and other herbivores. Proc. Natl. Acad. Sci. USA 85:511–514.

Chew, F.S. 1988. Biological effects of glucosinolates. *In* Biologically active natural products: potential use in agriculture, ed. H.G. Cutler, pp. 155–181. Am. Chem. Soc. Symp. Ser. No. 380. Washington, D.C.: American Chemical Society.

Choat, J.H., and Bellwood, D.R. 1985. Interactions amongst herbivorous fishes on a coral reef: influence of spatial variation. Mar. Biol. 89:221–234.

Cimino, G., Crispino, A., Di Marzo, V., Gavagnin, M., and Ross, J.D. 1990. Oxytoxins, bioactive molecules produced by the marine opishthobranch mollusc *Oxynoe olivacea* from a diet-derived precursor. Experientia 46:767–770.

Coen, L.D., and Tanner, C.E. 1989. Morphological variation and differential susceptibility to herbivory in the tropical brown alga *Lobophora variegata*. Mar. Ecol. Prog. Ser. 54:287–298.

Conn, E.E. 1979. Cyanide and cyanogenic glycosides. *In* Herbivores: their interaction with secondary plant metabolites, ed. G.A. Rosenthal and D.H. Janzen, pp. 387–412. New York: Academic Press.

Dart, J.K.G. 1972. Echinoids, algal lawn and coral recolonization. Nature 239:50–51.

Dawe, R.D., and Wright, J.L.C. 1986. The major polypropionate metabolites from the sacoglossan mollusc *Elysia chlorotica*. Tetrahedron Lett. 27:2559–2562.

Duffy, J.E., and Hay, M.E. 1990. Seaweed adaptations to herbivory. BioScience 40:368–375.

Earle, S.A. 1972. The influence of herbivores on the marine plants of Great Lameshur Bay, with an annotated list of plants. *In* Results of the tektite program: ecology of coral reef fishes, ed. B.B. Collette and S.A. Earle. Nat. Hist. Mus. Los Ang. Cty. Sci. Bull. 14:17–44.

Faulkner, D.J. 1984. Marine natural products: metabolites of marine algae and herbivorous marine molluscs. Nat. Prod. Rep. 1:251–280.

Faulkner, D.J. 1986. Marine natural products. Nat. Prod. Rep. 3:1–33.

Faulkner, D.J. 1987. Marine natural products. Nat. Prod. Rep. 4:539–576.

Faulkner, D.J. 1988a. Marine natural products. Nat. Prod. Rep. 5:613–663.

Faulkner, D.J. 1988b. Feeding deterrents in molluscs. *In* Biomedical importance of marine organisms, ed. D.G. Fautin, pp. 29–36. San Francisco: California Academy of Sciences.

Fenical, W. 1975. Halogenation in the Rhodophyta: a review. J. Phycol. 11:245–259.

Gaines, S.D. 1985. Herbivory and between-habitat diversity: the differential effectiveness of defenses in a marine plant. Ecology 66:473–485.

Gerwick, W.H., Fenical, W., and Norris, J.N. 1985. Chemical variation in the tropical seaweed *Stypopodium zonale* (Dictyotaceae). Phytochemistry 24:1279–1283.

Hackney, J.M., Carpenter, R.C., and Adey, W.H. 1989. Characteristic adaptations to grazing among algal turfs on a Caribbean coral reef. Phycologia 28:109–119.

Hatcher, B.G., and Larkum, A.W.D. 1983. An experimental analysis of factors controlling the standing crop of the epilithic algal community on a coral reef. J. Exp. Mar. Biol. Ecol. 69:61–84.

Hay, M.E. 1981a. Spatial patterns of grazing intensity on a Caribbean barrier reef: herbivory and algal distribution. Aquat. Bot. 11:97–109.

Hay, M.E. 1981b. Herbivory, algal distribution, and the maintenance of between habitat diversity on a tropical fringing reef. Am. Nat. 118:520–540.

Hay, M.E. 1984a. Patterns of fish and urchin grazing on Caribbean coral reefs: are previous results typical? Ecology 65:446–454.

Hay, M.E. 1984b. Predictable spatial escapes from herbivory: how do these affect the evolution of herbivore resistance in tropical marine communities? Oecologia 64:396–407.

Hay, M.E. 1985. Spatial patterns of herbivore impact and their importance in maintaining algal species richness. *In* Proc. Fifth Int. Coral Reef Symp., vol. 4, pp. 29–34.

Hay, M.E. 1986. Associational plant defenses and the maintenance of species diversity: turning competitors into accomplices. Am. Nat. 128:617–641.

Hay, M.E. 1991a. Marine-terrestrial contrasts in the ecology of plant chemical defenses against herbivores. Trends Ecol. Evol., in press.

Hay, M.E. 1991b. Fish-seaweed interactions on coral reefs: effects of herbivorous fishes and adaptations of their prey. *In* The ecology of fishes on coral reefs, ed. P. F. Sale, pp. 96–119. San Diego: Academic Press.

Hay, M.E., Colburn, T., and Downing, D. 1983. Spatial and temporal patterns in herbivory on a Caribbean fringing reef: the effects on plant distribution. Oecologia 58:299–308.

Hay, M.E., Duffy, J.E., and Fenical, W. 1988d. Seaweed chemical defenses: among-compound and among-herbivore variance. *In* Proc. Sixth Int. Coral Reef Symp., vol. 3., pp. 43–48.

Hay, M.E., Duffy, J.E., and Fenical, W. 1990a. Host-plant specialization decreases predation on a marine amphipod: an herbivore in plant's clothing. Ecology 71:733–743.

Hay, M.E., Duffy, J.E., Fenical, W., and Gustafson, K. 1988c. Chemical defense in the seaweed *Dictyopteris delicatula*: differential effects against reef fishes and amphipods. Mar. Ecol. Prog. Ser. 48:185–192.

Hay, M.E., Duffy, J.E., Paul, V.J., Renaud, P.E., and Fenical, W. 1990b. Specialist herbivores reduce their susceptibility to predation by feeding on the chemically-defended seaweed *Avrainvillea longicaulis*. Limnol. Oceanogr. 35:1734–1743.

Hay, M.E., Duffy, J.E., Pfister, C.A., and Fenical, W. 1987a. Chemical defense against different marine herbivores: are amphipods insect equivalents? Ecology 68:1567–1580.

Hay, M.E., and Fenical, W. 1988. Marine plant-herbivore interactions: the ecology of chemical defense. Annu. Rev. Ecol. Syst. 19:111–145.

Hay, M.E., Fenical, W., and Gustafson, K. 1987b. Chemical defense against diverse coral-reef herbivores. Ecology 68:1581–1591.

Hay, M.E., Paul, V.J., Lewis, S.M., Gustafson, K., Tucker, J., and Trindell, R.N. 1988a. Does the tropical seaweed *Halimeda* reduce herbivory by growing at night? Diel patterns of growth, nitrogen content, herbivory, and chemical versus morphological defenses. Oecologia 75:233–245.

Hay, M.E., Pawlik, J.R., Duffy, J.E., and Fenical, W. 1989. Seaweed-herbivore-predator interactions: host plant specialization reduces predation on small herbivores. Oecologia 81:418–427.

Hay, M.E., Renaud, P.E., and Fenical, W. 1988b. Large mobile versus small sedentary herbivores and their resistance to seaweed chemical defenses. Oecologia 75:246–252.

Hay, M.E., and Taylor, P.R. 1985. Competition between herbivorous fishes and urchins on Caribbean reefs. Oecologia 65:591–598.

Hiatt, R.W., and Strasburg, D.W. 1960. Ecological relationships of the fish fauna on coral reefs of the Marshall Islands. Ecol. Monogr. 30:65–127.

Hixon, M.A., and Brostoff, W.N. 1983. Damselfish as keystone species in reverse: intermediate disturbance and diversity of reef algae. Science 220:511–513.

Hobson, E.S. 1974. Feeding relationships of teleostean fishes on coral reefs in Kona, Hawaii. Fish. Bull. U.S. 72:915–1031.

Hodgson, L.M. 1984. Antimicrobial and antineoplastic activity in some south Florida seaweeds. Bot. Mar. 27:387–390.

Horn, M.H. 1989. Biology of marine herbivorous fishes. Oceanogr. Mar. Biol. Annu. Rev. 27:167–272.

Hughes, T.P., Reed, D.C., and Boyle, M.J. 1987. Herbivory on coral reefs: community structure following mass mortalities of sea urchins. J. Exp. Mar. Biol. Ecol. 113:39–59.

Huston, M.A. 1985. Patterns of species diversity on coral reefs. Annu. Rev. Ecol. Syst. 16:149–177.

Ireland, C., and Faulkner, D.J. 1981. The metabolites of the marine molluscs *Tridachiella diomedea* and *Tridachia crispata*. Tetrahedron 37:223–240.

John, D.M., and Pople, W. 1973. The fish grazing of rocky shore algae in the Gulf of Guinea. J. Exp. Mar. Biol. Ecol. 11:81–90.

Kami, H.T., and Ikehara, I.I. 1976. Notes on the annual juvenile siganid harvest in Guam. Micronesica 12:323–325.

Lagler, K.F., Bardach, J.E., Miller, R.R., and Passino, D.R.M. 1977. Ichthyology. New York: John Wiley and Sons.

Lawrence, J.M. 1975. On the relationships between marine plants and sea urchins. Oceanogr. Mar. Biol. Annu. Rev. 13:213–286.

Lessios, H.A. 1988. Mass mortality of *Diadema antillarum* in the Caribbean: what have we learned? Annu. Rev. Ecol. Syst. 19:371–393.

Lessios, H.A., Robertson, D.R., and Cubit, J.D. 1984. Spread of *Diadema* mass mortality through the Caribbean. Science 226:335–337.

Levitan, D.R. 1988. Algal-urchin biomass responses following mass mortality of *Diadema antillarum* Philippi at St. John, U.S. Virgin Islands. J. Exp. Mar. Biol. Ecol. 119:167–178.

Lewis, S.M. 1985. Herbivory on coral reefs: algal susceptibility to herbivorous fishes. Oecologia 65:370–375.

Lewis, S.M. 1986. The role of herbivorous fishes in the organization of a Caribbean reef community. Ecol. Monogr. 56:183–200.

Lewis, S.M., and Wainwright, P.C. 1985. Herbivore abundance and grazing intensity on a Caribbean coral reef. J. Exp. Mar. Biol. Ecol. 87:215–228.

Littler, M.M., and Littler, D.S. 1980. The evolution of thallus form and survival strategies in benthic marine macroalgae: field and laboratory tests of a functional form model. Am. Nat. 116:25–44.

Littler, M.M., Littler, D.S., and Taylor, P.R. 1987. Animal-plant defense associations: effects on the distribution and abundance of tropical reef macrophytes. J. Exp. Mar. Biol. Ecol. 105:107–121.

Littler, M.M., Taylor, P.R., and Littler, D.S. 1983. Algal resistance to herbivory on a Caribbean barrier reef. Coral Reefs 2:111–118.

Littler, M.M., Taylor, P.R., and Littler, D.S. 1986. Plant defense associations in the marine environment. Coral Reefs 5:63–71.

Lobel, P.S., and Ogden, J.C. 1981. Foraging by the herbivorous parrotfish *Sparisoma radians*. Mar. Biol. 64:173–183.

Lubchenco, J., and Gaines, S.D. 1981. A unified approach to marine plant-herbivore interactions. I. Populations and communities. Annu. Rev. Ecol. Syst. 12:405–437.

McConnell, O.J., Hughes, P.A., Targett, N.M., and Daley, J. 1982. Effects of secondary metabolites from marine algae on feeding by the sea urchin, *Lytechinus variegatus*. J. Chem. Ecol. 8:1437–1453.

McLachlan, J., and Craigie, J.S. 1966. Antialgal activity of some simple phenols. J. Phycol. 2:133–135.

Mathieson, A.C., Fralick, R.A., Burns, R., and Flashive, W. 1975. Phycological studies during Tektite II, at St. John U.S.V.I. *In* Results of the tektite program: coral reef invertebrates and plants, ed. S.A. Earle and R.J. Lavenberg. Nat. Hist. Mus. Los Ang. Cty. Sci. Bull. 20:77–103.

Meyer, K.D., and Paul, V.J. Submitted. Intraplant variation in secondary metabolite concentration in *Caulerpa* and its effects on herbivorous fishes. Mar. Ecol. Prog. Ser.

Morrison, D. 1988. Comparing fish and urchin grazing in shallow and deeper coral reef algal communities. Ecology 69:1367–1382.

Ogden, J.C. 1976. Some aspects of herbivore-plant relationships on Caribbean reefs and seagrass beds. Aquat. Bot. 2:103–116.

Ogden, J.C., Brown, R.A., and Salesky, N. 1973. Grazing by the echinoid *Diadema antillarum* Phillipi: formation of halos around West Indian patch reefs. Science 182:715–717.

Ogden, J.C., and Lobel, P.S. 1978. The role of herbivorous fish and urchins in coral reef communities. Environ. Biol. Fishes 3:49–63.

Paul, V.J. 1985. The natural products chemistry and chemical ecology of tropical green algae of the order Caulerpales. Ph.D. dissertation, University of California, San Diego.

Paul, V.J. 1987. Feeding deterrent effects of algal natural products. Bull. Mar. Sci. 41:514–522.

Paul, V.J., and Fenical, W. 1983. Isolation of halimedatrial: chemical defense adaptation in the calcareous reef-building alga *Halimeda*. Science 221:747–749.

Paul, V.J., and Fenical, W. 1986. Chemical defense in tropical green algae, order Caulerpales. Mar. Ecol. Prog. Ser. 34:157–169.

Paul, V.J., and Fenical, W. 1987. Natural products chemistry and chemical defense in tropical marine algae of the phylum Chlorophyta. *In* Bioorganic marine chemistry, ed. P.J. Scheuer, pp. 1–29. Berlin: Springer-Verlag.

Paul, V.J., and Hay, M.E. 1986. Seaweed susceptibility to herbivory: chemical and morphological correlates. Mar. Ecol. Prog. Ser. 33:255–264.

Paul, V.J., Hay, M.E., Duffy, J.E., Fenical, W., and Gustafson, K. 1987a. Chemical defense in the seaweed *Ochtodes secundiramea* (Rhodophyta): effects of its monoterpenoid components upon diverse coral-reef herbivores. J. Exp. Mar. Biol. Ecol. 114:249–260.

Paul, V.J., Littler, M.M., Littler, D.S., and Fenical, W. 1987b. Evidence for chemical defense in the tropical green alga *Caulerpa ashmeadii* (Caulerpaceae: Chlorophyta): isolation of new bioactive sesquiterpenoids. J. Chem. Ecol. 13:1171–1185.

Paul, V.J., Nelson, S.G., and Sanger, H.R. 1990. Feeding preferences of adult and juvenile rabbitfish *Siganus argenteus* in relation to chemical defenses of tropical seaweeds. Mar. Ecol. Prog. Ser. 60:23–34.

Paul, V.J., and Pennings, S.C. 1991. Diet-derived chemical defenses in the sea hare *Stylocheilus longicauda* (Quoy and Gaimard 1824). J. Exp. Mar. Biol. Ecol., in press.

Paul, V.J., and Potter, T.S. 1992. Seaweed susceptibility to herbivory on Guam reefs. J. Exp. Mar. Biol. Ecol., in press.

Paul, V.J., and Van Alstyne, K.L. 1988a. Chemical defense and chemical variation in the genus *Halimeda*. Coral Reefs 6:263–269.

Paul, V.J., and Van Alstyne, K.L. 1988b. The use of ingested algal diterpenoids by the ascoglossan opisthobranch *Elysia halimedae* Macnae as antipredator defenses. J. Exp. Mar. Biol. Ecol. 119:15–29.

Paul, V.J., and Van Alstyne, K.L. 1988c. Antiherbivore defenses in *Halimeda*. *In* Proc. Sixth Int. Coral Reef Symp., vol. 3, pp. 133–138.

Paul, V.J., and Van Alstyne, K.L. Submitted. Activation of chemical defenses in the tropical green algae *Halimeda* spp. J. Exp. Mar. Biol. Ecol.

Paul, V.J., Wylie, C., and Sanger, H. 1988. Chemical defenses of tropical seaweeds: effects against different coral-reef herbivorous fishes. *In* Proc. Sixth Int. Coral Reef Symp., vol. 3, pp. 73–78.

Pederson, M., Roomans, G.M., and Hofsten, A.V. 1980. Cell inclusions containing bromine in *Rhodomela confervoides* (Huds.) Lamour. and *Polysiphonia elongata* Harv. (Rhodophyta; Ceramiales). Phycologia 19:153–158.

Pfister, C.A., and Hay, M.E. 1988. Associational plant refuges: convergent patterns in marine and terrestrial communities result from differing mechanisms. Oecologia 77:118–129.

Randall, J.E. 1961. Overgrazing of algae by herbivorous marine fishes. Ecology 42:812.

Randall, J.E. 1965. Grazing effect on seagrasses by herbivorous marine fishes. Ecology 46:255–260.

Randall, J.E. 1967. Food habits of reef fishes of the West Indies. Stud. Trop. Oceanogr. (Miami) 5:655–847.

Randall, J.E. 1974. The effect of fishes on coral reefs. In Proc. Second Int. Coral Reef Symp., Brisbane, pp. 159–166.

Roussis, V., Pawlik, J.R., Hay, M.E., and Fenical, W. 1990. Secondary metabolites of the chemically rich ascoglossan Cyerce nigricans. Experientia 46:327–329.

Russ, G. 1984a. Distribution and abundance of herbivorous grazing fishes in the central Great Barrier Reef. I. Levels of variability across the entire continental shelf. Mar. Ecol. Prog. Ser. 20:23–34.

Russ, G. 1984b. Distribution and abundance of herbivorous grazing fishes in the central Great Barrier Reef. II. Patterns of zonation of mid-shelf and outershelf reefs. Mar. Ecol. Prog. Ser. 20:35–44.

Sale, P.F. 1980. The ecology of fishes on coral reefs. Oceanogr. Mar. Biol. Annu. Rev. 18:367–421.

Sale, P.F. 1988. Perception, pattern, chance and the structure of reef fish communities. Environ. Biol. Fishes 21:3–15.

Sale, P.F., and Douglas, W.A. 1984. Temporal variability in the community structure of fish on coral patch reefs and the relation of community structure to reef structure. Ecology 65:409–422.

Sammarco, P.W. 1980. Diadema and its relationship to coral spat mortality: grazing competition and biological disturbance. J. Exp. Mar. Biol. Ecol. 45:245–272.

Sammarco, P.W. 1982. Effects of grazing by Diadema antillarum Phillipi (Echinodermata: Echinoidea) on algal diversity and community structure. J. Exp. Mar. Biol. Ecol. 65:83–105.

Sammarco, P.W., Levinton, J., and Ogden, J.C. 1974. Grazing and control of coral reef community structure by Diadema antillarum: a preliminary study. J. Mar. Res. 32:47–53.

Scott, F.J., and Russ, G.R. 1987. Effects of grazing on species composition of the epilithic algal community on coral reefs of the central Great Barrier Reef. Mar. Ecol. Prog. Ser. 39:293–304.

Sieburth, J.M., and Conover, J.T. 1965. Sargassum tannin, an antibiotic which retards fouling. Nature 208:52–53.

Steinberg, P.D. 1985. Feeding preferences of Tegula funebralis and chemical defenses of marine brown algae. Ecol. Monogr. 55:333–349.

Steinberg, P.D. 1986. Chemical defenses and the susceptibility of tropical brown algae to herbivores. Oecologia 69:628–630.

Steinberg, P.D. 1989. Biogeographical variation in brown algal polyphenolics and other secondary metabolites: comparison between temperate Australasia and North America. Oecologia 78:373–382.

Steinberg, P.D., and Paul, V.J. 1990. Fish feeding and chemical defenses of tropical brown algae in Western Australia. Mar. Ecol. Prog. Ser. 58:253–259.

Steneck, R.S. 1986. The ecology of coralline algal crusts: convergent patterns and adaptive strategies. Annu. Rev. Ecol. Syst. 17:273–303.

Steneck, R.S. 1988. Herbivory on coral reefs: a synthesis. *In* Proc. Sixth Int. Coral Reef Symp., vol. 1, pp. 37–49.

Steneck, R.S., and Watling, L. 1982. Feeding capabilities and limitations of herbivorous molluscs: a functional group approach. Mar. Biol. 68:299–319.

Stephenson, W., and Searles, R.B. 1960. Experimental studies on the ecology of intertidal environments at Heron Island. Aust. J. Mar. Freshwater Res. 11:241–267.

Targett, N.M., Targett, T.E., Vrolijk, N.H., and Ogden, J.C. 1986. Effect of macrophyte secondary metabolites on feeding preferences of the herbivorous parrotfish. Mar. Biol. 92:141–148.

Taylor, P.R., Littler, M.M., and Littler, D.S. 1986. Escapes from herbivory in relation to the structure of mangrove island macroalgal communities. Oecologia 69:481–490.

Trench, R.K. 1975. Of "leaves that crawl": functional chloroplasts in animal cells. Symp. Soc. Exp. Biol. 29:229–265.

Trench, R.K. 1980. Uptake, retention and function of chloroplasts in animal cells. *In* Endocytobiology, endosymbiosis and cell biology, ed. W. Schwemmler and H.E.A. Schenk, pp. 703–727. Berlin: De Gruyter.

Tsuda, R.T., and P.G. Bryan. 1973. Food preferences of juvenile *Siganus rostratus* and *S. spinus* in Guam. Copeia 1973:604–606.

Tsuda, R.T., and Wray, F.O. 1977. Bibliography of marine benthic algae of Micronesia. Micronesica 13:85–120.

Van Alstyne, K.L. 1988. Herbivore grazing increases polyphenolic defenses in the intertidal brown alga *Fucus distichus*. Ecology 69:655–663.

Van Alstyne, K.L., and Paul, V.J. 1988. The role of secondary metabolites in marine ecological interactions. *In* Proc. Sixth Int. Coral Reef Symp., vol. 1, pp. 175–186.

Van Alstyne, K.L., and Paul, V.J. 1990. The biogeography of polyphenolic compounds in marine macroalgae: why don't tropical brown algae use temperate defenses against herbivorous fish? Oecologia 84:158–163.

Vine, P.J. 1974. The effects of algal grazing and aggressive behavior of the fishes *Pomacentrus lividus* and *Acanthurus sohal* on coral reef ecology. Mar. Biol. 24:131–136.

Wanders, J.B.C. 1977. The role of benthic algae in the shallow reef of Curaçao (Netherlands Antilles). III. The significance of grazing. Aquat. Bot. 3:357–390.

Wylie, C.R., and Paul, V.J. 1988. Feeding preferences of the surgeonfish *Zebrasoma flavescens* in relation to chemical defenses of tropical algae. Mar. Ecol. Prog. Ser. 45:23–32.

Young, D.N., Howard, B.M., and Fenical, W. 1980. Subcellular localization of brominated secondary metabolites in the red alga *Laurencia snyderae*. J. Phycol. 16:182–185.

Chapter 2
Geographical Variation in the Interaction between Marine Herbivores and Brown Algal Secondary Metabolites

PETER D. STEINBERG

Brown algae (division Phaeophyta) are the largest and most abundant benthic marine plants in temperate seas throughout the world (Kirkman 1984; Dayton 1985; Schiel and Foster 1986). They occur in great abundance in both the littoral and sublittoral zones and form some of the most productive plant communities in the world (Mann 1973, 1982). Some species of kelp (order Laminariales) reach lengths of more than 40 m (Abbott and Hollenberg 1976). Brown algae are also often long-lived; some have life spans of a decade or more (Dayton et al. 1984; Cousens 1985; Klinger and DeWreede 1988). Moreover, while many individual brown algae are shorter lived than this, patches of kelps or fucoids (order Fucales) often persist for years or decades (Dayton et al. 1984). For herbivores and other organisms associated with these plants, this persistence of patches may be just as important as the persistence of individuals. Brown algae, particularly plants in the orders Fucales and Dictyotales, are also among the most common seaweeds in the tropics.

Because of their large size, numerical and ecological (Dayton 1975) dominance, persistence, the canopies they form, and other characteristics, many brown algae are analogous to trees of terrestrial systems. In the terms of Feeny (1976) they are often the most "apparent" plants found in benthic communities on hard substrata. Like trees, they are fed on by a variety of invertebrate and vertebrate herbivores that can have major effects on their population dynamics, community structure, successional patterns, and distribution (Lawrence 1975; Lubchenco and Gaines 1981; Gaines and Lubchenco 1982; Hawkins and Hartnoll 1983).

Brown algae produce an assortment of secondary metabolites that can act as chemical deterrents against herbivores or other natural enemies (bacteria, epiphytes, etc.). The most common are polyphenolic compounds (Ragan and Glombitza 1986), again analogous to the polyphenols or tannins of trees and other woody terrestrial plants (Swain 1979). Brown algae also produce a variety of other, nonpolar, secondary compounds (Faulkner 1984, 1986, 1987). Both polyphenolics and nonpolar secondary metabolites deter or otherwise affect marine herbivores (Geiselman and McConnell 1981; Steinberg 1985, 1988; Hay et al. 1987a, 1987b, 1988b; Van Alstyne 1988).

The purposes of this chapter are threefold. The first is to review the research on the interaction between brown algal secondary compounds and marine herbivores. I take a biogeographical approach because much of the variation in these interactions falls along geographical lines.

Second, I compare these marine systems with analogous terrestrial systems. Polyphenolics have played a major role in studies of terrestrial plant-herbivore interactions (Feeny 1970, 1976; Fox 1981; Coley et al. 1985), but generalizations about their ecological roles have recently changed considerably (Bernays et al. 1989). Thus a comparison with marine systems seems timely. In general the predictions and assumptions of terrestrial models are not well supported by research on marine systems.

The final thrust of the chapter is an evolutionary overview of the interaction between brown algal secondary metabolites and marine herbivores. Much of the variation in the production of secondary metabolites in brown algae is correlated with transspecific (generic level or above) taxonomy or occurs at a large spatial scale. Thus the interaction between algae and their herbivores appears to be both strongly taxonomically constrained and also a function of large-scale biological and physical differences among different biogeographical regions.

Geographical Variation in Brown Algal Secondary Metabolites

The conspicuous brown algae in temperate seas belong to two orders: the Laminariales (kelps) and the Fucales (fucoids). A third order, the Dictyotales, are uncommon in colder seas but often abundant in warm temperate and tropical waters, as are some Fucales (particularly the family Sargassaceae). Kelps, with a few rare exceptions, do not occur in tropical waters. I focus on these three orders because they (1) contain the most abundant species of brown algae, (2) fit most closely the analogy with terrestrial trees (particularly the Fucales and Laminariales), (3) are the most chemically rich of the brown algae (Faulkner 1984, 1986), and (4) have been the focus of most of the chemical and chemical ecological studies done on the Phaeophyta.

Throughout this chapter I discuss two classes of compounds. The first group, the "phlorotannins" (Koch et al. 1980), consist of polymers of 1,3,5-trihydroxybenzene (phloroglucinol) (1). A typical low-molecular-weight phlorotannin from the brown alga *Fucus vesiculosus* is tetrafucol B (2).

Phlorotannins are the only polyphenolics described from brown algae. Like tannins from terrestrial plants, they have the ability to bind to proteins or other macromolecules (Ragan and Glombitza 1986); however, they are structurally different from the diverse group of tannins found in terrestrial plants (Haslam 1981). The chemistry and aspects of the biology of phlorotannins have recently been reviewed by Ragan and Glombitza (1986). The second group of compounds are more structurally diverse and loosely defined, but they are nonpolar (lipophilic) and generally smaller than polyphenolics. These nonpolar metabolites include terpenes, alkylated or prenylated phenolics, and acetogenins (reviewed by Fenical 1982; Faulkner 1986, 1987).

These two classes of compounds appear to be the primary options for chemical defense in brown algae. Nitrogenous metabolites such as alkaloids are essentially restricted to the blue-green algae (Hay and Fenical 1988). Halogenated hydrocarbons are found in low amounts (parts per million) in some brown algae (Gschwend et al. 1985) but occur at similar levels in both palatable and unpalatable algae (Gschwend et al. 1985) and are not known to deter herbivores. Compounds such as arsenosugars in kelp (Edmonds and Francesconi 1981a, 1981b) appear to be the products of reactions the algae use to detoxify absorbed arsenic (Cooney et al. 1978; Edmonds and Francesconi 1981a, 1981b) and hence are unlikely to play a primary role as defenses.

Temperate Seas

Perhaps the most striking pattern in the secondary metabolites of temperate brown algae is the difference in polyphenolic levels between plants in Australasia (Australia and New Zealand) and North America (Fig. 2.1; Steinberg 1989). Many North American algae contain very low levels of polyphenolics (0–2% by dry mass) as measured by the Folin-Denis assay (Ragan and Glombitza 1986; this colorimetric assay has been extensively used to quantify algal polyphenol content and is discussed below. Subsequent results, unless otherwise noted, are also from this assay.) This is especially true of kelps (Geiselman 1980; Steinberg 1985), the dominant plants in the sublittoral and lower littoral. Conversely, fucoids, which are most abundant in the littoral and upper sublittoral, are generally high (>4% dry mass) in polyphenol content (Geiselman 1980; Steinberg 1985; Van Alstyne 1988; Denton et al. 1990; Targett et al., submitted). This also appears true for the few species in the Dictyotales that have been analyzed (Steinberg 1985), although the Folin-Denis assay can be less reliable for this group. Other likely chemical defenses are not known from kelps and fucoids in North America (Faulkner 1984, 1986, 1987). The Dictyotales, most common in warmer temperate waters in North America, contain a variety of terpenoid or prenylated phenolic compounds (McEnroe et al. 1977; Fenical 1982; Pathirana and Anderson 1984; Hay et al. 1987b). Thus North American algae exhibit great variation in phenolic levels among species, with many species containing very low levels. This variation follows (although see Estes and Steinberg 1988 for an exception) a bathymetric (depth) gradient which corresponds to a taxonomic gradient (littoral fucoids are generally phenolic rich, sublittoral kelps are not).

Algal phenolic levels in temperate Australasia are on average much higher than in North America (Fig. 2.1; Steinberg 1989). Steinberg (1989) compared phenolic levels of 37 species of fucoids and kelps from temperate Australia and New Zealand with analogous data from 25 North American species (Steinberg 1984, 1985). Median phenolic levels of the Australasian plants were 6.20% (dry mass), compared with 1.33% for

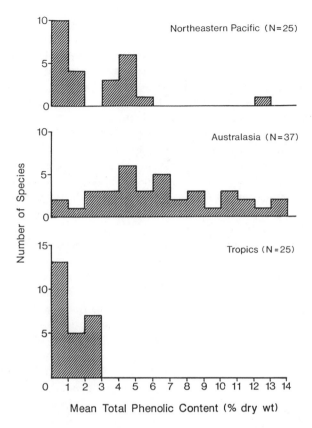

Figure 2.1 Mean phenolic levels, as analyzed by the Folin-Denis procedure, from species of kelps and fucoids from the temperate northeastern Pacific, temperate Australia and New Zealand, and assorted tropical Indo-Pacific sites (fucoids only). (Data from Steinberg 1984, 1985, 1986, 1989; Steinberg and Paul 1990; Van Alstyne and Paul 1990; and Steinberg et al. 1991. Histograms a and b from Steinberg 1989, courtesy of *Oecologia*.)

the North American species, and many of the most common species in Australasia contained more than 10% total phenolics. In part these differences can be attributed to the higher diversity (Womersley 1969) of fucoids (typically phenolic rich) and the lower diversity (Womersley 1967, 1981) of kelps (typically phenolic poor) in temperate Australasia than in North America. However, these taxonomic differences do not explain all the large regional differences in phenolic levels among the algae. For example, the kelp *Ecklonia radiata*, the most abundant alga in Australia (Kirkman 1984), is a phenolic-rich species, and phenolic levels in Australasian fucoids are often higher than in North American species (Steinberg 1985, 1989). Another difference between the two regions arises because only one species of fucoids or kelps (*Hormosira banksii*) is common above the low intertidal in Australasia (Dakin 1980). The Australasian data are thus almost exclusively from sublittoral or lower littoral algae. Given the bathymetric pattern discussed above, this pattern further exaggerates the difference in phenolic levels between Australasian and North American brown algae.

The second major difference between Australasia and North America is that many common fucoids in temperate Australasia contain nonpolyphenolic, nonpolar metabolites such as terpenes or prenylated phenolics (Kazlauskas et al. 1981; Gregson et al. 1982; van Altena 1988). Thus the secondary metabolites found in the dominant brown algae differ qualitatively as well as quantitatively between Australasia and North America.

Levels of polyphenolics in brown algae in other temperate regions have either been assayed by a variety of methods, making comparisons difficult, or they are simply less well known. Many of these data (particularly for Europe) have been summarized by Ragan and Glombitza (1986). In general, other temperate regions seem more similar to North America than to Australasia in having a mixture of phenolic-rich and phenolic-poor species. This appears to be true of South Africa (Anderson and Velimirov 1982; Tugwell and Branch 1989) and the northwestern Pacific (Katayama 1951; Estes and Steinberg 1988), and the European North Atlantic (Ragan and Glombitza 1986: table 16), where the bathymetric distribution of kelps and fucoids is similar to that of North America (Lewis 1964; Schonbeck and Norton 1978). Cold temperate waters in South America are quite depauperate in both kelps and fucoids (6–7 genera total; see Nizammuddin 1970; Santelices et al. 1980; Santelices and Ojeda 1984), all of which appear to contain low levels of polyphenolics (B. Santelices, pers. comm., 1988).

Unlike North America, fucoids in these other temperate regions also contain nonpolar secondary metabolites (Kato et al. 1975; Shizuru et al. 1982; Amico et al. 1985). The Dictyotales produce terpenoids and other secondary metabolites wherever they occur (McEnroe et al. 1977; Fenical 1982; Pathirana and Anderson 1984). Nonpolar secondary metabolites are unknown from kelps, with the exception of the recent isolation of ecklonia-lactones from the Japanese kelp *Ecklonia stolonifera* (Kurata et al. 1989).

Tropics

Most data on phenolic levels in tropical brown algae in the orders Fucales and Dictyotales are from analyses of Indo-Pacific species.

These analyses indicate consistently low levels of phenolics in these algae—typically ≤2.5% by dry mass (Fig 2.1; Steinberg 1986; Steinberg and Paul 1990; Van Alstyne and Paul 1990; Steinberg et al. 1991). In fact, since colorimetric assays such as the Folin-Denis test can respond to nonpolyphenolic but reactive compounds, polyphenolics may be completely absent from some tropical brown algae (Van Alstyne and Paul 1990). This difference in polyphenolic levels between tropical and temperate brown algae in the Indo-Pacific has been reported (Steinberg and Paul 1990; Van Alstyne and Paul 1990) as a rare example of a latitudinal gradient in the secondary metabolites of marine organisms in which levels are higher in temperate species than in tropical ones (in contrast to the patterns described by Bakus 1969; Green 1977; Vermeij 1978; Fenical 1980). The low levels of polyphenolics in brown algae in the tropical Indo-Pacific also contrasts with their production of nonpolar metabolites, which are abundant and diverse in tropical Dictyotales (Fenical 1982; Gerwick and Fenical 1982) and occasionally found in tropical Fucales.

However, the generality of this pattern of low phenolic levels in tropical brown algae is confounded by recent data from Caribbean species found near Belize. Mean levels of phenolics in four species of fucoids and two species of Dictyotales from Belize varied between 3.3 and 15.5% by dry weight (Targett et al., submitted). These levels are comparable to the highest levels found in temperate species. Moreover, one of the most phenolic-rich species (*Lobophora variegata*) studied by Targett et al. in Belize contains low levels of polyphenolics in western Australia (Steinberg and Paul 1990). We do not yet know if algae from other sites in the Caribbean (or other, as yet unstudied, tropical locations) also contain high levels of phenolics.

In summary, there seem to be three general patterns of secondary metabolite production described for brown algae from different regions of the world (Fig. 2.1, Table 2.1): (1) a "typical" cold temperate pattern, in which phenolic-rich fucoids are abundant in the littoral and shallow sublittoral zones (particularly in the Northern Hemisphere), and predominantly phenolic-poor kelps are common in the sublittoral; this pattern is modified by variation in the abundance of Dictyotales, which are rich in nonpolar metabolites, and the presence or absence of nonpolar metabolites in fucoids, particularly in the families Sargassaceae and Cystoseiraceae; (2) a temperate Australasian pattern, in which the algae are consistently rich in polyphenolics, and nonpolar metabolites are also common; and (3) a tropical Indo-Pacific pattern, in which polyphenolics are low or absent, and nonpolar compounds are common in the Dictyotales and infrequent in fucoids. A

58   Peter D. Steinberg

Table 2.1

Taxonomic patterns in the production of secondary metabolites by marine brown algae

	Levels of polyphenolics		
	Temperate	Tropical Indo-Pacific[a]	Nonpolar metabolites
Order			
Fucales	High	Low	Variable[b]
Laminariales	Low[c]	—	Not found[d]
Dictyotales	High(?)	Low	Common

[a]Phenolic levels in brown algae from Belize (Caribbean) are often high (Targett et al., submitted). Levels in algae from other Caribbean or tropical Atlantic sites are not known.

[b]Absent in the family Fucaceae; most common in the Sargassaceae and Cystoseiraceae.

[c]Exceptions include *Agarum, Ecklonia,* and *Dictyoneurum*.

[d]Except for *Ecklonia stolonifera* (Kurata et al. 1989)

Source: condensed from Faulkner 1984, 1986, 1987; Steinberg 1985, 1986, 1989; Ragan and Glombitza 1986; Hay and Fenical 1988; and Targett et al., submitted.

general pattern for polyphenolic levels in brown algae from other tropical regions has not yet been established (Targett et al., submitted).

Smaller-Scale Patterns

The patterns described above are generalizations, and variations within themes will occur. Individual species can consistently differ from the overall trend, such as the common phenolic-rich kelp *Agarum cribrosum* in the Northern Hemisphere (Estes and Steinberg 1988), or the phenolic-poor *Macrocystis* spp. in Australasia. Kelps are sometimes absent altogether from warmer temperate communities (Pielou 1977), and the fucoid flora of South Africa is quite different (Nizamuddin 1970) from the typical intertidal bands of *Fucus, Pelvetia,* and so forth found in the Northern Hemisphere.

Moreover, substantial intraspecific or intraplant variation in levels or kinds of secondary metabolites occurs in many species of brown algae (Geiselman 1980; Steinberg 1984; Gerwick et al. 1985; Johnson and Mann 1986; Ragan and Glombitza 1986; Van Alstyne 1988; Tugwell and Branch 1989; Denton et al. 1990), and this variation can be ecologically important (Steinberg 1984; Johnson and Mann 1986; Van Alstyne 1988). Ragan and Glombitza (1986) have provided a detailed review of smaller-scale patterns of phenolic variation in brown algae. Consistent environmental or other correlates of intraspecific or intraplant variation in phenolic levels are difficult to discern, but at least two patterns seem to emerge: (1) variation in phenolic levels within a species appears to be positively correlated with salinity (Pederson 1984), and (2) meristodermal or cortical tissues in the Laminariales are generally richer in phenolics than inner medullary tissues (Ragan and Glombitza 1986; Tugwell and Branch 1989; Steinberg, pers. observ.). Seasonal variation occurs, but it seems to be species or population specific (Ragan and Glombitza 1986: table 17; cf. Ragan and Jensen 1978, and Geiselman 1980). The phenolic content of juvenile plants may be higher (Steinberg 1989) or lower (Denton et al. 1990) than that in adult plants of the same species. Reproductive tissues may also have higher (Steinberg 1984) or lower (Steinberg 1989) phenolic levels than vegetative tissues. Although there are consistent biogeographical and higher-order taxonomic patterns in the production of secondary metabolites by brown algae, intraplant or intraspecific variation can be considerable, and it is wise to recall Janzen's (1979) caution that herbivores do not eat species or genera; they eat individual plants or plant parts.

Qualitative (Structural) Variation in Secondary Metabolites

Structural variation in brown algal polyphenolics, and its ecological consequences, is not well known. In part this is due to the technical problems of isolating large, structurally similar polar molecules. The structures of a variety of individual polyphenolic molecules from brown algae have been described, but these are exclusively low-molecular-weight oligomers (Ragan and Glombitza 1986; although see Ragan 1985), which are often not the major fraction. The proportion of polymers of different molecular weights is known for some species (Ragan 1976; Geiselman and McConnell 1981). Some information on structural variation in polyphenolics can also be derived from correlations with taxonomy. For example, "fucols" (polymers in which phloroglucinol units are linked only through aryl-aryl bonds) are known only from the Fucales (Ragan and Glombitza 1986); thus spatial or temporal variation in these compounds will reflect that of the Fucales.

For the immediate future, our understanding of patterns of variation in algal polyphenolics is likely to continue to be based primarily on colorimetric assays such as the Folin-Denis test. Although the Folin-Denis assay has been used extensively to assay levels of phenolics in both terrestrial and marine plants (e.g., Swain and Hillis 1959; Ragan and Jensen 1977; J. S. Martin and Martin 1982; reviews by Tempel 1982; Ragan and Glombitza 1986; Mole and Waterman 1987; Hagerman and Butler 1989), it does not provide information about the specific kinds of phenolic molecules present in a sample. It is also subject to the same difficulties of any colorimetric assay, including variation in extractability of the target compounds, variation in reactivity of different phenolics (probably less of a problem for many algal phenolics than for terrestrial compounds; Steinberg 1988), and the presence of non-phenolic but reactive compounds (e.g., ascorbic acid). Van Alstyne and Paul (1990) suggested that the Folin-Denis test may not be sensitive enough to distinguish between algae with very low levels of phenolics and those lacking phenolics completely. However, more precise (i.e., nuclear magnetic resonance) analyses of typical phenolic-poor taxa such as *Laminaria, Alaria,* and tropical *Sargassum* species do generally indicate the presence of polyphenolics (Ragan and Glombitza 1986; Steinberg et al. 1991). Other colorimetric assays such as vanillin-H_2SO_4 (Lindt's reagant) may be even less sensitive to phenolic-poor extracts than the Folin-Denis assay (Ragan and Glombitza 1986). The reliability of the Folin-Denis assay is discussed in the publications cited above and in Ragan and Jensen (1977) and Steinberg (1985, 1988).

Since nonpolar metabolites have been analyzed primarily through the isolation and identification of pure compounds, considerably more is known about the exact structure of these metabolites than is known for polyphenolics (Faulkner 1984, 1986, 1987). Until recently, however, the focus in these studies was largely at the species level. Thus, while we know a fair amount about the kinds of compounds in different species or genera, studies of intraspecific variation of algal nonpolar metabolites are still rare (Gerwick et al. 1985; Paul and Van Alstyne 1988). Because even small differences in molecular structure among nonpolar algal metabolites can cause large differences in their biological effects (McConnell et al. 1982; Hay et al. 1987b, 1988c), this sort of variation is ecologically important and warrants further study.

Responses of Marine Herbivores to Secondary Metabolites

Temperate North America

Most studies on the interaction between temperate herbivores and brown algal secondary metabolites have used North American species. They show that polyphenolics consistently deter feeding by a range of herbivores, as do selected nonpolar metabolites, and the data strongly suggest that these compounds can enhance the fitness of algae in the field.

In experimental tests, algal phenolics incorporated into palatable agar disks significantly deterred feeding by gastropod or echinoid herbivores (Geiselman and McConnell 1981; Steinberg 1988). At concentrations of 0.001–0.1% fresh weight, polyphenolics from *Ascophyllum nodosum* and *Fucus vesiculosus*, phloroglucinol, and the terrestrial polyphenol tannic (polygallic) acid all decreased feeding by the gastropod *Littorina littorea* by at least 50% (Geiselman and McConnell 1981). Deterrence varied somewhat among different molecular weight fractions, suggesting both qualitative and quantitative differences among polyphenolics. The monomer, phloroglucinol, exhibited effects similar to the polyphenolics, although the authors failed to take into account leaching of phloroglucinol out of their agar disks (Steinberg 1988). Geiselman and McConnell's (1981) results correlate well with the generally low palatability of these fucoid algae to *Littorina* in New England (Lubchenco 1978, 1983).

Steinberg's (1988) results were similar. Feeding rates of the gastropods *Tegula funebralis* and *T. brunnea* were decreased by 50–70% by polyphenols from *F. vesiculosus*, *Halidrys siliquosa*, and *Eisenia arborea*, and tannic acid, added at concentrations of 0.2–0.5% fresh weight to palatable agar disks (not all compounds were tested against both herbivores). Phloroglucinol and tannic acid deterred feeding by the echinoid *Strongylocentrotus purpuratus*. Phloroglucinol had no effect on *T. funebralis*. Steinberg (1988) argued that because of the similarity of the deterrent effects of the different polyphenolics (excluding phloroglucinol), the effects of algal polyphenolics on these herbivores were primarily quantitative (dosage) rather than a function of qualitative differences between different molecules. This conclusion is further supported by the recent demonstration of feeding deterrence of *T. funebralis* by polyphenolics from three Australasian algae, *Ecklonia radiata*, *Sargassum vestitum*, and *Carpophyllum maschalocarpum* (Steinberg and van Altena 1992). Steinberg's (1988) conclusion is tempered, however, by the fact that in these experiments a "mixed polyphenol" fraction was used from each algal species. Possible differences in the effects of specific phenolic polymers are obscured by these techniques.

The generally low palatability of phenolic-rich algae to invertebrate herbivores in North America is strong indirect evidence for the deterrent effects of polyphenols. Steinberg's (1984, 1985) laboratory and field data for *T. funebralis*, and summary results for 17 other species of invertebrate herbivores from the northeastern Pacific, indicated consistent preferences by these herbivores for phenolic-poor brown algae over phenolic-rich species, individuals, or tissues. Similar correlations have been found by Johnson and Mann (1986), Estes and Steinberg (1988), Van Alstyne (1988), Barker and Chapman (1990), and Denton et al. (1990). An exception to this pattern is the chiton *Cyanoplax hartwegii*, which in some habitats feeds exclusively on *Pelvetia fastigiata*

(Robb 1975), a phenolic-rich species. No analogous data are available for fishes, but gut contents of northeastern Pacific herbivorous fishes such as the opaleye (*Girella nigricans*) and halfmoon (*Medialuna californiensis*) suggest a phenolic-poor diet (Quast 1968).

While polyphenolics have clear behavioral effects (feeding deterrence) on North American herbivores, much less is known about their physiological effects. Estes and Steinberg (1988) summarized data from several studies that showed echinoids grew more slowly and produced less gonadal tissue when fed phenolic-rich algae than when fed phenolic-poor species. Assimilation efficiencies of phenolic-rich algae also are often less than those of phenolic-poor algae (Lowe and Lawrence 1976; Vadas 1977; Larson et al. 1980). One exception to this trend was shown by Keats et al. (1984), who found that the urchin *Strongylocentrotus drobachiensis* grew fairly rapidly when fed *Fucus vesiculosus*, a typically phenolic-rich species. Phenolic levels in *Fucus vesiculosus* vary seasonally (Ragan and Jensen 1978; Geiselman 1980), however, and these urchins may have been feeding on relatively phenolic-poor plants.

Even less data exist for herbivorous fishes. The assimilation efficiency of nitrogen by the stichaeid fish *Cebidichthys violaceus* is less when fed the phenolic-rich *Fucus distichus* than when fed the phenolic-poor kelp *Macrocystis integrifolia* (Horn et al. 1985). Neither alga is typically eaten by *C. violaceus*, however, and assimilation efficiency of both species was low, suggesting factors other than polyphenolics were involved (Horn et al. 1985).

The effects of nonpolar compounds from brown algae on North American marine herbivores have been studied by Hay and his colleagues (Hay et al. 1987b, 1988a) in North Carolina. In these studies preference for *Dictyota menstrualis* was very low in the echinoid herbivore *Arbacia punctulata* and two herbivorous fishes (*Diplodus holbrooki* and *Lagodon rhomboides*). *D. menstrualis* contains two cyclic diterpene alchohols, pachydictyol-A (3) and dictyol-E (4)

3   4

both of which significantly deterred feeding by the two fishes when coated on otherwise palatable algae. Hay et al. (1987b) also showed that pachydictyol-A inhibits the growth of *Diplodus holbrooki*, one of the few direct experimental demonstrations of a physiological effect of algal secondary compounds on a marine herbivore. *Arbacia punctulata* is deterred by dictyol-E but not by pachydictyol-A (Hay et al. 1987b). Smaller mesograzers such as amphipods and polychaetes readily consume *Dictyota menstrualis* and are not deterred from feeding by either dictyol-E or pachydictyol-A (Hay et al. 1987b, 1988a). This supports the hypothesis (Hay et al. 1987b) that the effects of secondary compounds on smaller, more sedentary herbivores will be different from their effects on large, mobile herbivores such as fishes or sea urchins. Mesograzers may be more tolerant to polyphenolics as well.

Several field studies have demonstrated increased survival of phenolic-rich species, plants, or parts of plants, indicating that phenolics can function as chemical defenses under natural conditions. At the interspecific level, grazing on phenolic-rich intertidal fucoids in New England is generally low relative to other more ephemeral algae (Lubchenco 1978, 1982, 1983). These plants often dominate habitats where herbivory is high, although the importance of herbivory varies considerably among habitats (Lubchenco 1983). There is also variation in the interaction between different herbivores and different species of fucoids in the northwestern Atlantic (Barker and Chapman 1990). In sublittoral communities in the Aleutian Islands, the phenolic-rich kelp *Agarum cribrosum* is less susceptible to grazing than phenolic-poor *Laminaria* spp., and *A. cribrosum* typically dominates Aleutian habitats where herbivores are abundant (Estes and Steinberg 1988 and unpublished).

At the intraspecific level, simulated herbivory (clipping) on *Fucus distichus* in Washington State increased phenolic levels, relative to unclipped plants, by ≈20% (Van Alstyne 1988). Grazing by *Littorina sitkana* and *L. scutulata* on clipped plants in the field was subsequently decreased by ≈50%. This is the only evidence for induction of chemical defenses in a marine alga (Van Alstyne 1988), and it may explain the differences in phenolic levels among populations of *F. distichus* observed by Van Alstyne (1988). Ecologically important intraplant variation in phenolic levels also occurs in the kelps *Alaria marginata* (Steinberg 1984) and *Laminaria saccharina* (Johnson and Mann 1986). Higher phenolic content in the sporophylls of *A. marginata* and the meristem of *L. saccharina* are both correlated with decreased herbivory on these tissues relative to more phenolic-poor parts of the algae. Both studies are consistent with the hypothesis (McKey 1979) that plants allocate production of secondary compounds in a way that maximizes protection of those parts or tissues most crucial to future fitness.

The generality of the deterrent effects of polyphenolics in North

America should be qualified, because with a few exceptions (Johnson and Mann 1986; Estes and Steinberg 1988) the data are from studies of littoral gastropods (Geiselman 1980; Geiselman and McConnell 1981; Steinberg 1984, 1985, 1988; Van Alstyne 1988). Phenolic-rich species are consistently unpalatable to echinoids as well as molluscs (Vadas 1977; Steinberg 1985), some phenolics deter feeding by echinoids in the laboratory (Steinberg 1988), and phenolic-rich algae can inhibit the growth of echinoids (above). Estes and Steinberg (1988) argued that the deterrent effects of polyphenolics on echinoids have played an important role in the evolution of the kelp *Agarum cribrosum* and in determining the zonation patterns of sublittoral kelps in North America (Tremblay and Chapman 1980). The extent to which herbivorous echinoids feed selectively in the field, however, and thus their effect on the community structure of temperate brown algae (Schiel and Foster 1986), is controversial. In many instances "urchin fronts" may remove all or nearly all fleshy algae (Harrold and Pearse 1987) in sublittoral kelp forests; alternatively, echinoids may subsist largely on drift, thereby having little effect on living plants (Harrold and Reed 1985). In both instances polyphenolics would not be functioning as defenses. It would be interesting to know if plants in the Dictyotales or *Cystoseira* spp. persisted in areas of moderate urchin grazing. Both taxa are rich in polyphenolics or nonpolar metabolites and are found interspersed with phenolic-poor kelps in the northeastern Pacific.

Temperate Australasia

Interactions between marine herbivores and the chemical defenses of brown algae in temperate Australasia are much less well known than in North America. The available evidence indicates that the responses of Australasian marine herbivores to brown algal secondary metabolites differ significantly from those of North American herbivores.

Common herbivorous echinoids and gastropods in Australasia, and some fishes, are generally not deterred from feeding on algae rich in polyphenolics. For example, there is no correlation between Schiel's (1982) data on feeding preferences of the echinoid *Evechinus chloroticus* in northern New Zealand and variation in phenolic levels in the common brown algae there (Steinberg 1989). Steinberg and van Altena (1992) found a similar lack of correlation between feeding preferences of *E. chloroticus* and the sublittoral gastropod *Cookia sulcata*, and variation in phenolic levels among seven species of laminarian and fucoid brown algae from New Zealand. Phenolic levels in brown algae from the Sydney region are also not consistently correlated with the feeding preferences of the common echinoids *Centrostephanus rodgersii*, *Tripneustes gratilla*, and *Holopneustes pycnotilus* (Steinberg and van Altena 1992). The gastropod *Turbo undulata* in New South Wales, however, is deterred from feeding on some phenolic-rich algae, but not others, indicating that some herbivores in Australasia do show sensitivity to certain phenolics. Fishes in temperate Australasia can also be quite tolerant of phenolic-rich algae. For example, the labroid fishes *Odax pullus* and *O. cyanomelas* in northern New Zealand and southern Australia respectively feed largely on *Ecklonia radiata* or *Carpophyllum* spp. (Clements 1985; Andrew and Jones 1990), which typically contain 5–15% total phenolics (Steinberg 1989).

Increased tolerance to polyphenolics by Australasian invertebrate herbivores has been confirmed experimentally by Steinberg and van Altena (1992) using the agar disk techniques of Geiselman and McConnell (1981) and Steinberg (1988). At concentrations (5 mg/ml in the agar disks) that consistently deterred North American herbivores (Steinberg 1988), polyphenolics extracted from the kelp *Ecklonia radiata* and the fucoids *Sargassum vestitum* and *Carpophyllum maschalocarpum* failed to deter feeding by the echinoid *Centrostephanus rodgersii* in New South Wales. At these same concentrations the Australian gastropod *Turbo undulata* was undeterred by polyphenolics from *S. vestitum*, *Carpophyllum maschalocarpum*, and the North American algae *Fucus vesiculosus* (fucoid) and *Dictyoneurum californicum* (kelp), but was significantly deterred by compounds from *E. radiata*. *Tripneustes gratilla* from New South Wales was undeterred by polyphenolics from *E. radiata*, *D. californicum*, *C. maschalocarpum*, and *F. vesiculosus*, but was deterred by compounds from *S. vestitum*. In northern New Zealand the gastropods *Turbo smaragdus* and *Cookia sulcata*, and the echinoid *Evechinus chloroticus*, are generally undeterred by phenolics from the fucoid algae *Carpophyllum maschalocarpum* and *F. vesiculosus*, but are deterred by compounds extracted from *E. radiata* (P. D. Steinberg and J. A. Estes, unpubl. data).

When compared with analogous work in North America (Geiselman and McConnell 1981; Steinberg 1988), these results indicate a much higher frequency and level of tolerance to algal phenolics among Australasian herbivores. But the interaction between marine herbivores and polyphenols is also more complex in Australasia. Different polyphenols can have different effects on a given herbivore (as for the effects of *E. radiata* compounds versus fucoid phenolics in New Zealand), and polyphenolics from one species of algae can have variable effects on different herbivores (as for the effects of *E. radiata* phenolics on New Zealand versus Australian herbivores). While this complexity in the effects of algal metabolites is perhaps not surprising (Hay and Fenical 1988), the deterrence of New Zealand herbivores by polyphenolics from *E. radiata* is somewhat paradoxical. *E. radiata* is a preferred food of at least one of these herbivores (the echinoid *E. chloroticus*; Schiel 1982). Furthermore, it is removed from subtidal boulders at a rapid rate by echinoids (Schiel 1982), and it also supports a higher rate of gonadal development in *E. chloroticus* than do other species of algae (Andrew 1986).

The longer-term effects of polyphenolics on Australasian herbivores appear to be much weaker than on North American herbivores. In New South Wales, the growth of juvenile *Tripneustes gratilla* (echinoids) grown on agar disks was not significantly inhibited by the addition of 5 mg/ml of polyphenols from *S. vestitum* over six months (during which the urchins more than doubled in size; Steinberg and van Altena, in press). Growth of these urchins over seven months was also not correlated with variation in algal phenolic content (Steinberg and van Altena 1992). Similar levels of polyphenolics from *S. vestitum* and *E. radiata* also did not significantly affect growth of *Turbo undulata* over eight months. Over the same period, however, *T. undulata* did grow significantly better on a relatively (for Australasia) phenolic-poor species of *Sargassum* than on several other more phenolic-rich species of brown algae, again suggesting that these snails may be affected by phenolics at high levels.

There is no clear evidence that polyphenolics function as natural defenses against Australasian herbivores. Algal phenolic levels are not correlated with the rate of removal of plants from subtidal boulders by *Evechinus chloroticus* in northern New Zealand (Schiel 1982; Steinberg 1989). In New South Wales, the most important effects of many limpets and echinoids occur at very young stages of the plants (Fletcher 1987; Jones and Andrew 1990), probably before polyphenolics could have an effect. *Holopneustes pycnotilus*, an urchin that feeds on adult plants in New South Wales, can affect the growth of adult *Ecklonia radiata* (kelp), but the severity of this effect is not correlated with variation in phenolic levels among individual plants or parts of plants (pers. observ.). In these latter studies I also failed to find any evidence for the induction of elevated levels of phenolics in *Ecklonia* due to grazing by *H. pycnotilus*. *Ecklonia* polyphenolics are also not effective deterrents against the herbivorous fish *Odax cyanomelas*, which clears patches in *Ecklonia* beds in New South Wales (Andrew and Jones 1990).

The effects of nonpolar compounds on Australian herbivores are even less well known. The Australasian fucoid flora is particularly rich in nonpolar metabolites, particularly in the family Cystoseiraceae

(Gregson et al. 1982; Kazlauskas et al. 1981; van Altena 1988). Algae that contain nonpolar metabolites are generally less preferred by the Australian echinoids *Tripneustes gratilla, Holopneustes pycnotilus,* and *Centrostephanus rodgersii* (Steinberg and van Altena 1990), and the terpene geranylacetone extracted from *Cystophora moniliformis* deters feeding by *T. gratilla.* Steinberg and van Altena (1992) hypothesized that nonpolar compounds more frequently act as deterrents against herbivores in Australasia than do polyphenolics. So far, too few compounds and too few herbivores have been investigated to test this hypothesis adequately.

Other Temperate Regions

Very little work on the responses of temperate herbivores to algal chemical defenses has been done outside North America and Australasia. The few existing studies suggest a pattern of interaction similar to that in North America. The echinoid *Parechinus angulosus* in South Africa prefers phenolic-poor brown algae over phenolic-rich species (Anderson and Velimirov 1982). In Scotland, several species of fucoids are generally low in palatability to *Littorina littorea* (Watson and Norton 1985). Even the germlings of these fucoids are unpalatable (Watson and Norton 1985), suggesting that chemical defenses may sometimes be effective even in very small plants.

Tropics

Levels of polyphenolics in the Indo-Pacific tropical algae analyzed so far are probably too low to have any substantial effect on herbivores, and thus polyphenolics may not be relevant to the ecology of these systems. Why don't these tropical algae produce higher levels of phenolics, given (1) the importance of phenolics as chemical defenses in some temperate systems and (2) the clear ability of some tropical algae to produce high levels of phenolics? One hypothesis is that tropical Indo-Pacific herbivores are not deterred or otherwise affected by phenolic-rich algae, and thus there has been no selection for high levels of polyphenolics in tropical algae.

Two studies have addressed this hypothesis. Van Alstyne and Paul (1990) coated methanolic extracts (containing polyphenolics) of temperate algae on blades of palatable tropical algae and presented these to herbivorous fishes in Guam. Extracts of phenolic-rich species such as *Fucus distichus* or *Agarum fimbriatum* consistently deterred feeding by the fishes; extracts from phenolic-poor species did not. Steinberg et al. (1991), however, found that phenolic-poor tropical species of *Sargassum* were no more susceptible to grazing by herbivorous fishes than phenolic-rich temperate species on the Great Barrier Reef. The fishes were not deterred by phenolic-rich algae. As in temperate systems, the effects of phenolics on tropical herbivores can vary considerably among different locales, although this conclusion is tempered by the possible confounding effects of methodological differences between the studies of Steinberg et al. (1991) and Van Alstyne and Paul (1990). In light of these studies, it would be of considerable interest to know how herbivores in Belize respond to the phenolic-rich algae found there (Targett et al., submitted).

Brown algae that contain nonpolar compounds are often less palatable to tropical herbivorous fishes or echinoids, and are among the most avoided species in field assays of herbivory (Littler et al. 1983, 1986; Hay 1984; Hay et al. 1987a). In field experimental tests, terpenes such as dictyol-E (4) and pachydictyol-A (3) from species in the Dictyotales deterred specific herbivores or made the plants less susceptible to mixed assemblages of herbivores (Hay 1987a; Wylie and Paul 1988). Tropical fishes are also deterred by the C-11 hydrocarbons dictyopterenes A and B (5)

5

from the Caribbean alga *Dictyopteris delicatula* (Hay et al. 1988b). As with temperate herbivores, mesograzers are generally much more tolerant to these compounds from tropical algae (Hay et al. 1987a, 1988b; Hay this volume) than are large, mobile herbivores such as fishes or echinoids. One tropical amphipod, *Pseudoamphithoides incurvaria*, in fact uses chemicals that deter feeding by reef fishes as cues for selection of an appropriate habitat (Hay et al. 1990). Tropical dictyotalean algae are rich in these compounds (Fenical 1982; Gerwick et al. 1985) and are often among the most prominent brown algae on tropical coral reefs.

Although data are still scant for most areas of the world, some general patterns of effects of brown algal secondary metabolites on large, mobile marine herbivores seem to be emerging. In North America, algal phenolic levels are variable, and herbivores are frequently deterred by phenolic-rich algae. Phenolics consistently act as chemical defenses, although there is a taxonomic and spatial bias in these studies (gastropods, usually in the littoral zone). Nonpolar metabolites from the Dictyotales are also deterrent. In temperate Australasia, herbivores show considerable tolerance to even high levels of polyphenolics, there is variation in the response of herbivores to different poly-phenolics, and polyphenolics may not be particularly important as chemical defenses against herbivores. Nonpolar compounds, which are more common in Australian than in North American algae, may be more frequently deterrent and may function as ecologically important chemical defenses. Other temperate areas seem more similar to North America than to Australasia. In the tropics, levels of polyphenolics are often, but not always (Targett et al., submitted), very low and have not been shown to be ecologically important. Nonpolar compounds, par-ticularly in the Dictyotales, are common chemical defenses.

Comparison with Terrestrial Plant-Herbivore Systems

Tannins

Tannins (polyphenols that bind protein) have probably been the single most important class of compounds in the development of ter-restrial plant-herbivore theory during the past 20 years. Research on polyphenolics in marine systems has been strongly influenced by these ideas (Geiselman and McConnell 1981; Steinberg 1985; Ragan and Glombitza 1986), and thus a comparison between the effects on herbivores of terrestrial tannins versus marine phlorotannins seems appropriate.

The presumed importance of tannins as chemical defenses in ter-restrial plants is based on several observations or hypotheses. Initially, it was thought that these compounds (1) were the dominant class of defensive compounds in trees and other "apparent" plants (Feeny 1976; Rhoades and Cates 1976; Swain 1977); (2) functioned (in a physi-ological sense) primarily through binding to plant protein or the diges-tive enzymes of herbivores, thereby reducing the assimilation of nutri-ents—particularly nitrogen—by herbivores (Feeny 1968, 1969); (3) were "quantitative" defenses, in the sense that as their concentration in the plant increased, they were increasingly effective against herbivores (Feeny 1976; Rhoades and Cates 1976); (4) were difficult for herbivores to adapt to, unlike many other secondary compounds (Rhoades 1979); and (5) were costly to the plant in terms of the amount of energy or other resources that had to be expended in order to produce and maintain them (Rhoades 1979; Coley et al. 1985).

The recent review by Bernays et al. (1989) indicates considerable

changes in our view of tannins as defenses against terrestrial herbivores. First, while tannins may be the dominant defensive compounds in apparent plants, with clear effects on the behavior and
physiology of many invertebrate and vertebrate herbivores, they are a
very diverse group of compounds. Different tannins undoubtedly function in different ways (Zucker 1983; Bernays et al. 1989), and grouping,
for example, condensed and hydrolyzable tannins into one general
category obscures considerable chemical and functional information
(Zucker 1983).

Second, while these compounds have deleterious physiological effects on many herbivores, their main mechanism of action is probably
not the reduction of digestibility by binding to plant proteins or the
digestive enzymes of the herbivore (J. S. Martin et al. 1987; Blytt et al.
1988; Bernays et al. 1989). In insects, tannins probably act mainly by
causing lesions in the gut epithelium, with subsequent deleterious effects on internal tissues or organs (Bernays et al. 1981, 1989; Berenbaum 1984). In vertebrates, a variety of mechanisms (Mehansho et al.
1987; Blytt et al. 1988) prevent tannins from binding digestive enzymes.
Plant protein may sometimes be complexed by tannins (Hagerman
and Robbins 1987), but other effects are probably more important (Bernays et al. 1989). These effects include ulceration of the gut with subsequent damage to internal organs (Robbins et al. 1987), inhibition of
microbe activity in ruminants (Cooper and Owen-Smith 1985), and dis-
ruption of mineral balance (Freeland et al. 1985b).

Third, tannins do not differ from other secondary metabolites in
their quantitative nature (Fox 1981). Most biologically active chemicals
generally show some dosage effect. Thus, in a general sense tannins
appear to be "toxins," and not fundamentally different from a wide
variety of other secondary metabolites. A consideration of tannins as
toxins also eliminates a potential paradox of digestibility-reducing
compounds: Why don't the herbivores simply eat more?

Fourth, since their mode of action is not fundamentally different
from other secondary metabolites, it is not surprising that mechanisms for coping with dietary tannins are common among both vertebrate and invertebrate herbivores. Insects minimize the effects of tannins by producing surfactants (M. M. Martin and Martin 1984), raising
the gut pH (Berenbaum 1980), and thickening the gut membranes (Bernays et al. 1981, 1989). Vertebrates also produce or ingest surfactants
(Freeland et al. 1985a), and increased mucus production is probably a
common mechanism for minimizing tannins' effects (Freeland et al.
1985b). Many vertebrates inactivate tannins through the production of
salivary proteins that bind to tannins before they enter the gut

(Mehansho et al. 1987). Additional mechanisms are probable. Like insects that feed on plants containing other secondary compounds, herbivorous insects that commonly feed on tanniferous plants are less affected by tannins than herbivores that do not (Bernays et al. 1989: table 1).

From the plant's point of view as well, tannins may be no different from other carbon-based secondary metabolites. They function as both constitutive and inducible defenses (Feeny 1970; Schultz and Baldwin 1982), are responsive to environmental and internal physiological changes (McLure 1979), and may be no more costly to produce than other compounds (Swain 1978; Fox 1981; below). Bernays et al. (1989), while emphasizing that tannins have a wide variety of deleterious effects on terrestrial herbivores, suggest as a "working hypothesis" that tannins may have evolved primarily as defenses against fungi and pathogens (also see Zucker 1983).

Considerably less is known about algal polyphenolics, but their ecological effects seem similar to those described above. They deter feeding by many herbivores but have little effect against others. Little is known about the physiological effects of algal polyphenolics on marine herbivores. Tanniferous algae in North America inhibit the growth of herbivores (above), but the mechanism of action is not known. Long-term effects of polyphenolics on Australasian herbivores seem less strong, and thus, as in terrestrial systems, the effects of polyphenolics appear weaker on herbivores that commonly feed on phenolic-rich plants (as in Australasia). These herbivores probably possess mechanisms that inactivate polyphenolics. Because the taxa of herbivores in marine systems (echinoids, fishes) are quite different from those which feed on terrestrial plants, the specifics of these mechanisms may be different from those used by terrestrial herbivores.

As in terrestrial plants (Zucker 1983), algal polyphenolics also have other functions besides deterrence of herbivores. These compounds may inhibit bacterial or fungal growth and infection (Sieburth 1968), inhibit epiphytes (Conover and Sieburth 1964; Sieburth and Conover 1965; although see Anderson and Velimirov 1982; Ronneberg and Ruokolahti 1986), or act as a sink for excess carbon (e.g., Bryant et al. 1983). Some other proposed functions seem less likely. For example, algal polyphenolics are unlikely to function as screens against ultraviolet radiation, since they absorb most strongly at wavelengths (Ragan and Craigie 1980; Ragan and Glombitza 1986) that are shorter than most incident ultraviolet radiation (Smith and Baker 1979; Dunlap et al 1986). Lignin is not known from algae (Grisebach 1981; although Ragan and Glombitza [1986] discuss equivocal evidence for the presence of lignin in *Fucus*), and hence algal polyphenolics are also unlikely to act as intermediates in lignin synthesis.

There are at least two other important differences and one similarity between terrestrial and marine polyphenolics that deserve brief mention here. A major difference is that the compounds are different. Polyphloroglucinols are not found in vascular terrestrial plants (Swain 1979; Haslam 1981); condensed and hydrolyzable tannins are not found in marine algae (Ragan and Glombitza 1986). The biosynthetic pathways of these compounds are also different (Ragan and Glombitza 1986:176–177; W. Fenical, pers. comm., 1989). These structural differences may reflect functional differences as well. The second difference is that while terrestrial plants in various locales may typically contain high levels of tannins (Macauley and Fox 1980), the magnitude of biogeographical variation in phenolic levels in terrestrial plants seems less than that described for brown algae (although see Coley and Aide 1990).

The similarity is that in both marine and terrestrial plants, polyphenolics (and phenolics in general) are usually produced in much higher concentrations (20–30% dry mass in some cases) than other secondary metabolites. The levels needed to affect herbivores are also often higher than for other secondary metabolites (Steinberg 1988; Hay et al. 1987a). This is an intriguing parallel, but its significance is not yet clear.

Evolutionary Models: Plant Apparency and Resource Availability

One of the major goals of evolutionary ecology is to synthesize present-day ecological patterns into a consistent evolutionary framework. A number of such models for the evolution of plant-herbivore interactions have been developed for terrestrial communities. These models are based on very general characteristics of plants and herbivores and should be applicable to marine plant-herbivore interactions as well.

The two best known general models for the evolution of defensive characteristics of plants and the responses of their herbivores are the plant apparency model developed by Feeny (1976) and Rhoades and Cates (1976; also see Rhoades 1979), and the resource availability model (Bryant et al. 1983; Coley et al. 1985). These models have been very useful and influential in directing research. The plant apparency model assumes that large or persistent plants are "bound to be found" by herbivores, and as a consequence must invest heavily in dosage-dependent, or "quantitative," defenses such as polyphenols or morphological toughness, which will deter or at least inhibit most herbivores. Unapparent, ephemeral plants, such as herbs and early successionals, will avoid most herbivory in space and time. They are predicted to produce toxins such as alkaloids or cyanogenic compounds, which deter common generalist herbivores. Most herbivory on these plants will be due to specialists that have adapted to these particular toxins.

The resource availability model (Coley et al. 1985) makes similar predictions in a number of instances but focuses more strongly on the plants. Allocation to defensive properties is tied strongly to the potential growth rate of a plant and the resources available to the plant. For slow-growing plants in nutrient-poor environments, any loss to herbivory is very costly, and thus investment in defense against herbivores should be high. When resources are abundant and growth rates are high, losses to herbivores are relatively easy to replace, and investment in defenses should result in a greater absolute reduction in growth rates than in slow-growing plants. Thus investment in defenses by fast-growing plants in nutrient-rich habitats should be low. Finally, as the persistence of plant parts (leaves in the case of trees) increases, mobile chemical defenses such as alkaloids, cyanogenic compounds, and so forth should be replaced by immobile defenses such as tannins or morphological defenses, which have high initial construction costs but low maintenance costs (Coley et al. 1985).

Both models seem applicable for both between-group (e.g., trees versus herbs) and within-group (among trees) comparisons. The two models differ in that the resource availability model makes no explicit predictions about the effects on, or responses of, different types of herbivores (e.g., specialists and generalists; Coley et al. 1985) whereas this is a crucial component of the plant apparency model (Feeny 1976). Because brown algae vary greatly in many of the important parameters of these models (apparency, type of defense, growth rate, availability of nutrients in their habitat), they would seem to be an ideal group in which to test these models.

The plant apparency model does not explain the observed variation in secondary metabolites among brown algae. Apparent plants may contain very low (*Macrocystis, Laminaria, Nereocystis*, many tropical fucoids) or very high (*Ecklonia radiata*, most temperate fucoids) levels of polyphenolics. The presence of other secondary compounds seems independent of the life history or apparency of the algae, and largely a function of taxonomy. Thus species of *Dictyota* in the tropics and in temperate regions contain similar sorts of nonpolar metabolites (Pathirana and Anderson 1984), even though the regimes of herbivory faced and the general ecological milieus are quite different.

One difficulty in using apparency theory to predict patterns in the production of algal secondary metabolites production is the rarity of ecologically important specialist herbivores in marine systems (Lubchenco and Gaines 1981; Hay and Fenical 1988; Hay, this volume). Almost all examples of herbivores that affect algal populations involve fishes, echinoids, or gastropods, which feed on a wide variety of algae. One of the fundamental components of the plant apparency model is largely missing from marine systems. The resource availability model is more difficult to test with available data because fairly detailed knowledge of growth rates of the plants and rates of herbivory may be required and such data are largely unavailable. It is generally thought, however, that herbivory is higher on coral reefs (Gaines and Lubchenco 1982) than in temperate kelp forests, resources (nutrients) are generally less available on reefs than in temperate upwelling systems, and incident light is higher in the tropics. If this is the case, then tropical brown algae should be rich in polyphenolics—exactly the opposite of what is often observed. Other carbon-based, nonpolar metabolites found commonly in temperate fucoids (Kato et al. 1975; Kazlauskas et al. 1981; Shizuru et al. 1982) also appear to be uncommon in related tropical species. Brown algae in similar temperate systems in Australasia and North America, both of which occur in relatively nutrient-rich water, differ dramatically in the levels of secondary compounds they produce.

Some support for the resource availability model comes from a comparison among different taxa of temperate algae. Some fucoids and the Dictyotales grow slowly compared with kelps, which are among the fastest-growing plants in the world (Mann 1973). Fucoids and Dictyotales are generally richer in secondary metabolites than kelps. Within the kelps, *Agarum cribrosum*, a phenolic-rich species, grows more slowly than phenolic-poor *Laminaria* spp., although there is considerable overlap in growth rates (Mann 1972; Estes and Steinberg 1988). But there are many important exceptions. *Sargassum* spp. and members of the Cystoseiraceae can grow extremely rapidly (Norton 1977; DeWreede 1979; Schiel 1985) but are as phenolic rich or richer than other fucoids (Steinberg 1985, 1989), and also contain other metabolites (Faulkner 1984, 1986). *Ecklonia radiata* in Australia contains high levels of phenolics, but the growth rate (stipe elongation) of this species is slightly greater than the related, phenolic-poor North American species *Pterygophora californica* (DeWreede 1984; Novaczek 1984).

The two major classes of secondary compounds in brown algae—polyphenols and smaller, nonpolar molecules—are analogous (or homologous for some nonpolar compounds) to the dichotomies between quantitative and qualitative (Feeny 1976), or immobile versus mobile (Coley et al. 1985) defenses. Neither the apparency model nor the resource availability model, however, explains the distribution of different kinds of secondary metabolites in brown algae. First, there appears to be no inverse correlation between levels of polyphenolics produced and the production of nonpolar compounds in brown algae (Steinberg 1989), as one would expect from the resource availability model. Second, fast-growing brown algae do not, as a rule, produce higher levels of the smaller, nonpolar, mobile compounds. In fast-growing taxa these compounds are either absent (in the Laminariales, except for *Ecklonia stolonifera*; Kurata et al. 1989) or occur only sporadically (*Sargassum*, Cystoseiraceae). They are most common in slower-growing taxa such as the Dictyotales. Again, much of the pattern of variation of these compounds among brown algae is taxonomically based. Thus cyclic diterpenes (Fenical 1982) are the "characteristic" metabolites of the Dictyotaceae (Faulkner 1986) but are rare or absent in other brown algal taxa.

The meaning of mobility versus immobility is also unclear with regard to algae, which lack xylem and phloem. Although translocation of nutrients (sugars, amino acids) through specialized cells does occur in the Laminariales and Fucales (Schmitz and Lobban 1976; Moss 1983), there is no evidence that secondary metabolites are moved throughout the thallus in these plants. Finally, the relevance of a *functional* distinction between polyphenolics (quantitative or immobile defenses) and smaller, nonpolar molecules (terpenes, etc.) in brown algae is unclear for either the plants or the herbivores (Hay and Fenical 1988; below).

Cost

These theories may be difficult to apply to brown algae because of the difficulty of assessing what role the cost of secondary metabolite production has played in the evolution of these compounds. The optimization hypothesis—that secondary metabolites are costly to produce—is crucial to most recent theories for the evolution of plant defenses (Feeny 1976; Rhoades and Cates 1976; McKey 1979; Rhoades 1979; Coley et al. 1985; Hay and Fenical 1988; Van Alstyne and Paul 1988). It appears, however, that either the cost of production of secondary metabolites (particularly polyphenolics) in brown algae is relatively low, or these costs are manifested in indirect or diffuse ways.

First, there is little evidence that the "relative" costs of different types of compounds differ (Fox 1981; Hay and Fenical 1988). Although polyphenolics are generally produced in much greater quantities than nonpolar compounds, they are relatively inert metabolically and are stored in vesicles (physodes; Ragan 1976). Thus, while their initial "construction" costs may be high, their "maintenance" costs are low (Swain 1978; Fox 1981; Coley et al. 1985). The opposite may be true for smaller, nonpolar metabolites (Coley et al. 1985). It is not easy to assess where, over the lifetime of an alga, these two costs should balance out or cross. Moreover, since all these compounds are carbon based, and brown algae are probably limited by nutrients other than carbon (Chapman and Craigie 1977; Lobban et al. 1985), their production may not represent a substantial investment by the plants (Hay and Fenical 1988). In fact, the levels at which polyphenolics occur suggest that they may be quite cheap to produce—a poor plant's chemical defense. Differences in the cost of different but structurally related compounds—which may constitute less than 1% of the mass of an alga but can vary enormously in their effects on herbivores (McConnell et al. 1982; Hay et al. 1987b)—are even more difficult to evaluate. For example, the addition of a hydroxyl group to a diterpene can change the molecule's effectiveness as a deterrent (Hay et al. 1987a, 1987b), but it seems unlikely that this minor chemical change represents a significant cost in resources to an alga.

Second, it is not clear what the "absolute" costs—the decrease in fitness caused by the production of secondary compounds in the absence of herbivory—of secondary compound production are to brown algae. As discussed above, a number of the correlations predicted between growth rate and secondary compounds do not seem to hold true for brown algae. If secondary compounds impose a significant cost on the algae, one would expect to see consistent decreases in production of secondary metabolites by well-defended taxa occurring in low herbivory habitats. While we cannot yet fully evaluate this hypothesis, the overall consistency of secondary metabolite production in many brown algae in the face of dramatically different environmental conditions suggests this does not occur. Widely distributed temperate genera or families such as *Sargassum*, the Cystoseiraceae, and *Fucus* consistently produce high levels of phenolics wherever they occur, even though the selective pressure from herbivores must vary enormously throughout their distribution. Similarly, the kelp genera *Macrocystis* and *Ecklonia* appear to be consistently phenolic poor or phenolic rich (respectively), even though they are widely distributed on at least three different continents. Species in the Dictyotales in tropical habitats contain metabolites similar to those found in temperate North America (Pathirana and Anderson 1984), where herbivory is almost certainly much less. Even in situations in which herbivores may be having a selective effect (Van Alstyne 1988) on the algae, the change in phenolic levels is from phenolic rich to very phenolic rich. Thus levels of phenolics in a population of *Fucus distichus* subjected to low rates of herbivory (studied by Van Alstyne 1988) are still comparable to levels found in many other populations of *F. distichus* (Steinberg 1985). Other algae, such as the green algal genus *Caulerpa*, are chemically rich in both low herbivory (sandflats) and high herbivory (reef slopes) habitats, although the relative proportions of different metabolites produced vary among these habitats (Paul and Fenical 1986; Paul and Van Alstyne 1988; V. J. Paul, pers. comm., 1989). While there is ecologically important variation in metabolite production in all these instances, the tendency for a given taxon to produce relatively consistent levels or kinds of compounds seems surprisingly strong.

From a more general perspective, the causality underlying cost arguments may not necessarily imply that the production of secondary compounds is costly. There seems to be considerable evidence in terrestrial systems to support the model outlined by Bryant et al. (1983) on the relationship between nutrient availability and carbon-based secondary metabolites such as polyphenolics. There is indirect support for this model in marine algae as well (Ilvessalo and Tuomi 1989). When nutrients (e.g., nitrogen) are limiting, photosynthesis is limited less than growth and excess carbon is turned into secondary metabolites. When nutrients are available, growth is emphasized and production of secondary metabolites is less. The causality in this model implies that major changes in resource allocation by the plant revolve around whether to increase growth. If growth can't be increased, excess carbon is turned into something potentially useful. This logic suggests that the plant is not growing slowly because it is producing high levels of secondary metabolites; rather, it is producing high levels of metabolites because it is growing slowly. When rapid growth is possible, fewer metabolites are produced, presumably because any excess photosynthate is channeled into growth. The model implies that changes in secondary metabolites are a secondary consequence of the allocation of resources to growth, not vice versa, and that growth is "costly" to the plant, but production of secondary metabolites may not be.

Phylogenetic Constraints and Spatial Variation in Herbivory

While the models outlined above may predict many of the patterns in terrestrial plant-herbivore interactions, neither the predicted patterns nor the underlying assumptions seem to fit the interaction between brown algae and their herbivores. Doubts regarding some aspects of these models have also been raised for terrestrial plants and herbivores (Fox 1981; Bernays et al. 1989), and for marine algae in general (Hay and Fenical 1988). I believe these models are inadequate for brown algae because they fail to account for the uniqueness of different taxa of herbivores and plants, or for major biological and physical differences among different geographical regions. Since biogeographical and taxonomic patterns in the relevant plants, herbivores, and chemicals in these marine systems are so striking, these larger-scale effects have likely been very important in the systems' evolution. I suggest that a model based on variation in the intensity of herbivory, coupled with phylogenetic constraints, may be useful in explaining much of the evolution of secondary metabolites, particularly polyphenolics, in brown algae. The interaction of these factors generates the following, admittedly speculative, model.

With respect to sublittoral habitats in North America, Estes and Steinberg (1988) argued that most species in the order Laminariales (kelps) diversified in habitats in which herbivory was low, and selection for production of secondary compounds was thus also low. The exception to this, the phenolic-rich genus *Agarum*, probably evolved in deeper waters where herbivory was more intense (Estes and Steinberg 1988). Since the only chemical deterrents known from kelps in North America are polyphenolics, the chemical responses of *Agarum* were limited.

In the intertidal in North America (and the Northern Hemisphere generally), selective feeding by gastropods or other herbivores clearly has an important effect on the structure of macroalgal communities (Lubchenco 1978, 1982, 1983; Sousa 1979; Lubchenco and Gaines 1981; Hawkins and Hartnoll 1983; Watson and Norton 1985; although see Barker and Chapman 1990), and it seems reasonable that the family Fucaceae evolved in this habitat, in which there was selection for the production of high levels of polyphenolics as defenses against herbivores. Other kinds of likely chemical defenses are not known from this family, so again the potential responses of the plants were limited.

Presumably the large, fleshy thallus found in both the Laminariales and Fucales, which provides a morphological obstacle to many herbivores (Steneck and Watling 1982; Padilla 1985), was also an important factor in the evolution of these two groups. Moreover, these two orders alone possess morphological structures for the translocation of nutrients (Schmitz and Lobban 1976; Moss 1983). If internal allocation of nutrients has played a role in the evolution of chemical defenses in algae, then this latter characteristic could have been very important in these groups.

Estes and Steinberg (1988) suggested that historically, herbivory in temperate Australasia may have been higher than in temperate North America. They proposed that this was due to the absence in Australasia of sea otters, highly effective predators of large invertebrate herbivores. The apparently greater diversity of herbivorous fishes in cold waters in Australasia (Choat 1982), including some endemic taxa that feed largely on brown algae (Clements 1985), lends some additional support for this hypothesis. Thus selection by herbivores for high levels of polyphenolics may have historically been stronger on subtidal algae in temperate Australasia than in North America.

In the intertidal zones of southeastern Australia and northern New Zealand, gastropods often remove most or all foliose macroalgae (Underwood 1980; Underwood and Jernakoff 1981; Creese 1988) rather than selecting some species over others, as seems typical in the Northern Hemisphere. Thus these molluscan grazers in Australasia may preclude the evolution of chemical defenses in intertidal algae by indiscriminately removing the algae from the rock when they are very small, presumably before chemical defenses are effective (although see Watson and Norton 1985). Only one species of fucoid or kelp is typically found above the low intertidal in Australasia (*Hormosira banksii*; Dakin 1980; Womersley 1981).

While the abundance of endemic chemically rich taxa such as *Carpophyllum* and *Cystophora* in Australasia potentially confounds such an adaptive scenario, the facts that, first, *Ecklonia radiata*, the most abundant alga in temperate Australsia, is phenolic rich even though it is a member of a generally phenolic-poor group (kelps); and, second, fucoids in Australasia often contain even higher levels of phenolics than "phenolic-rich" fucoids in North America, suggest that high levels of phenolics in Australia are not simply a historical accident but have been selected for. Given the tolerance to polyphenolics shown by many herbivores in Australasia (particularly the ecologically important echinoids), however, the continued production of consistently high levels of phenolics in these algae becomes something of a paradox. Such high phenolic levels may be due to (1) selective herbivory by as-yet-unstudied herbivores, (2) strong phylogenetic inertia or a relative lack of cost of the compounds, or (3) other factors that select for high levels of polyphenolics in Australian algae. From the herbivores' perspective, since so much of the flora is rich in phenolics, one would expect selection for tolerance to these compounds. Conversely, since the production of nonpolar compounds is quite variable among the algae, selection for adaptation to these compounds should be less intense, and they should be more frequently effective as deterrents.

The work of Targett et al. (submitted) in Belize make the patterns in the production of polyphenolics by tropical brown algae perplexing from several perspectives, since it indicates that tropical algae are indeed capable of producing high levels of phenolics. Moreover, the effects of phenolics against tropical herbivores are not consistent. Some tropical herbivores are deterred by phenolic-rich extracts (Van Alstyne and Paul 1990); some are not deterred by phenolic-rich algae (Steinberg et al. 1991). Variation in the intensity of herbivory within and among tropical habitats is considerable (Hay 1981, 1984; Hay et al. 1983; Russ 1984a, 1984b; Scott and Russ 1987), and the resultant variation in the selective effects of herbivory may in part explain the variation in phenolic levels among tropical algae. The fact that some tropical algae can produce high levels of phenolics would seem to rule out the hypothesis of Steinberg and Paul (1990) and Steinberg et al. (1991), who proposed that phenolic production in tropical algae was constrained by a lack of trace metals that serve as cofactors for biosynthetic enzymes (Mayer and Harel 1979). Clearly we need to know whether the results of Targett et al. are unique to Belize, and what effects the phenolic-rich algae in Belize have on the local herbivores.

The above scenario highlights a number of additional future directions. First, measurement of the spatial variation in herbivory at several spatial scales, as has been done on some coral reefs (Hay et al. 1983; Hay 1984; Scott and Russ 1987), is critical. Is herbivory more intense in the littoral than in the sublittoral in North America? Is it more intense in temperate Australasia than in North America?

Second, we need a clearer understanding of the evolutionary history of the algae. The fact that much of the variation in metabolites occurs at higher taxonomic or biogeographical scales implies macroevolutionary events. Thus we must treat the evolution of these algae and their herbivores as a macroevolutionary problem. We need to know more about the paleontological history of the relevant organisms, either through direct investigation of the fossil record (Steneck 1983) or through indirect means (Estes and Steinberg 1988).

Relating phylogenetic relationships within particular groups of algae to patterns in their production of secondary metabolites, as has been done in studies of other presumed adaptive traits (Ridley 1983; Coddington 1988), would be very useful. This should be facilitated by the increasing use of the tools of molecular biology in the analysis of relationships among the algae (Fain et al. 1988). Cladograms for chemically rich taxa are badly needed here because they would enable us to begin to assess which metabolites are primitive or derived—a crucial issue for understanding the evolution of algal chemical defenses. For example, the Laminariales probably evolved from Scytosiphonales-like algae (Wynne and Loiseaux 1976), apparently a relatively phenolic-poor group (pers. observ.), and *Laminaria*, the presumed primitive type of kelp (Druehl 1980), is low in phenolics. Thus high phenolic levels in kelps such as *Agarum* and *Ecklonia* are probably derived traits, and the prevalence of low phenolic levels in most kelps can be explained as the retention of a primitive character. If the converse were true (primitve kelps were rich in phenolics), then any evolutionary scenario would have to explain the loss of high phenolic levels in most of the kelp species. Similar comparisons between independently constructed cladograms (Coddington 1988) and patterns of secondary metabolites in chemically rich groups such as the Dictyotales or Caulerpales should prove enlightening. Is the diversity of compounds in these groups due largely to the appearance of new compounds being produced by new taxa? Are most kinds of metabolites conserved in new taxa? Are compounds lost in taxa that invade habitats where herbivory is low?

Third, and related to these issues, is the relative contribution of genotypic and phenotypic factors to variation in secondary metabolite production. How heritable are phenolics or other compounds? The patterns described above suggest that heritability is fairly high and that environmental effects are less important than taxonomic ones. Intraspecific variation can be high in some groups (Paul and Van Alstyne 1988). How much of this is genetic?

Fourth, if we want to evaluate the cost of secondary metabolites to the algae, we need to be able to distinguish among situations where (1) there is a real cost to the fitness of the plant; (2) compounds are not costly but are not particularly beneficial either, and so have not been selected for; and (3) metabolites are phylogenetically or physically constrained. We also must be able to demonstrate that the causality in models of cost works both ways. Although I have suggested here that costs are often low or absent, cost may be crucial to the production of metabolites in some algae, and at some level cost becomes relevant to all plants. Perhaps an appropriate taxon with which to test cost hypotheses is the Fucales, whose monophasic life history facilitates the sorts of selection experiments that are necessary (Berenbaum et al. 1986; Sims and Rausher 1987).

Conclusions

Unlike most other algal secondary compounds, polyphenolics in brown algae are largely a temperate phenomenon. Temperate algae now face, and have faced, a very different regime of herbivory (among many other factors) than that faced by tropical species. The dominant algae and herbivores at different latitudes are often taxonomically or morphologically very different, in some cases fundamentally so (e.g., the coenocytic Caulerpales versus the kelps). It may be that the rules governing these interactions in temperate systems are quite different from those in tropical systems.

Clear spatial and taxonomic patterns in secondary metabolite production exist for many other algae, however, and the ideas presented in this chapter may apply to other taxa as well. For example, the patterns of metabolite production described by Hay (1984) could be explained by intense selection for chemical defenses in high herbivory habitats, coupled with taxonomic and physical constraints on algae in low herbivory habitats. Thus in low herbivory habitats some algae will (e.g., *Caulerpa*) or won't (e.g., *Hypnea, Acanthophora*) produce chemical defenses, depending on their phylogenetic history or physical constraints. Finer-scale variation (Gerwick et al. 1985) might then be a function of differences in the kinds of metabolites selected *for* in different habitats (by herbivores or other factors) coupled with a healthy dose of biochemical noise.

Although I have emphasized broad-scale patterns in the interaction between marine herbivores and the secondary metabolites of brown algae, I do not deny the importance of smaller-scale variation. For example, while the Dictyotales were used as an example of a group that produces "nonpolar" compounds, the diversity of these compounds is considerable (Fenical 1982), and different compounds can have very different effects (Hay et al. 1987a, 1987b). The factors underlying the production of these different variations on a theme are very important. It is also important, however, to understand why many of the dominant brown algae in the world do not produce any nonpolar metabolites at all, or why so many algal taxa produce only one or a few main types of compounds wherever they are found. These patterns may not fit easily into models based solely on local ecological events, and they suggest that much of the variation that we see on ecological scales may not be important on evolutionary ones. At any rate, the variation in secondary compound production by algae at these larger scales must be incorporated into our models for the evolution of these interactions.

Finally, it is worth mentioning again that all secondary metabolites may have multiple functions. Herbivory is doubtless an important factor, but the effects of these compounds against bacteria and epiphytes or their function as allelopathic agents (de Nys et al. 1991) are largely unexplored in natural settings, and these interacting functions may have parallel or conflicting effects on the evolution of these compounds.

Acknowledgments I thank J. Estes, L. Fox, and V. Paul for their insightful comments. During the writing of this chapter I was supported by a Queen Elizabeth II Fellowship from the Australian government, an Australian Research Council Grant, and the School of Biological Sciences, University of Sydney.

References

Abbott, I.A., and Hollenberg, G.H. 1976. Marine algae of California. Stanford: Stanford University Press.

Amico, W., Cunsolo, F., Piatelli, M., and Ruberto, G. 1985. Acyclic tetraprenyltoluquinols from *Cystoseira sauvageuana* and their possible role as biogenetic precursors of the cyclic *Cystoseira* metabolites. Phytochemistry 24:2663–2668.

Anderson, R.J., and Velimirov, B. 1982. An experimental investigation of the palatability of kelp bed algae to the sea urchin *Parechinus angulosus*. Mar. Ecol. (Pubbl. Stn. Zool. Napoli I) 3:357–373.

Andrew, N.L. 1986. The interaction between diet and density in influencing reproductive effort in the echinoid *Evechinus chloroticus* (Val.). J. Exp. Mar. Biol. Ecol. 97:63–79.

Andrew, N.L., and Jones, G.P. 1990. Patch formation by herbivorous fish in a temperate Australian kelp forest. Oecologia 85:57–68.

Bakus, G.J. 1969. Energetics and feeding in shallow marine waters. Int. Rev. Gen. Exp. Zool. 4:275–369.

Barker, K.M., and Chapman, A.R.O. 1990. Feeding perferences among four species of *Fucus*. Mar. Biol. 107:113–118.

Berenbaum, M. 1980. Adaptive significance of midgut pH in larval Lepidoptera. Am. Nat. 115:138–146.

Berenbaum, M. 1984. Effects of tannins on growth and digestion in two species of papilionids. Entomol. Exp. Appl. 34:245–250.

Berenbaum, M.R., Zangerl, A.R., and Nitao, J.K. 1986. Constraints on chemical coevolution: wild parsnips and the parsnip webworm. Evolution 40:1215–1228.

Bernays, E.A., Chamberlain, D., and Leather, E.M. 1981. Tolerance of acridids to ingested condensed tannin. J. Chem. Ecol. 7:247–256.

Bernays, E.A., Cooper Driver, G., and Bilgener, M. 1989. Herbivores and plant tannins. Adv. Ecol. Res. 19:263–302.

Blytt, H.J., Guscar, T.K., and Butler, L.G. 1988. Antinutritional effects and ecological significance of dietary condensed tannins may not be due to binding and inhibiting digestive enzymes. J. Chem. Ecol. 14:1455–1465.

Bryant, J.P., Chapin, F.S., III, and Klein, D.R. 1983. Carbon/nutrient balance of boreal plants in relation to vertebrate herbivory. Oikos 40:357–368.

Chapman, A.R.O., and Craigie, J.S. 1977. Seasonal growth in *Laminaria longicruis*: relations with dissolved inorganic nutrients and internal reserves of nitrogen. Mar. Biol. 40:197–205.

Choat, J.H. 1982. Fish feeding and the structure of benthic communities in temperate waters. Annu. Rev. Ecol. Syst. 13:423–449.

Clements, K.D. 1985. Feeding in two New Zealand herbivorous fish, the butterfish *Odax pullus* and the marblefish *Aplodactylus arctidens*. M.Sc. thesis, University of Auckland.

Coddington, J.A. 1988. Cladistic tests of adaptational hypotheses. Cladistics 4:3–22.

Coley, P.D., and Aide, T.M. 1990. Comparison of herbivory and plant defenses in temperate and tropical broad-leaved forests. *In* Plant-animal interactions: evolutionary ecology in tropical and temperate regions, ed. P.W. Price, T.M. Lewinson, G.W. Fernandes, and W.W. Benson, pp. 25–49, John Wiley and Sons.

Coley, P.D., Bryant, J.P., and Chapin, F.S., III. 1985. Resource availability and plant antiherbivore defense. Science 230:895–899.

Conover, J.T., and Sieburth, J.M. 1964. Effect of *Sargassum* distribution on its epibiota and antibacterial activity. Bot. Mar. 6:147–157.

Cooney, R.V., Mumma, R.O., and Benson, A.R. 1978. Arsoniumphospholipid in algae. Proc. Natl. Acad. Sci. USA 75:4262–4264.

Cooper, S.M., and Owen-Smith, N. 1985. Condensed tannins deter feeding by browsing ungulates in a South African savanna. Oecologia 67:142–146.

Cousens, R. 1985. Frond size distributions and the effects of algal canopy on the behaviour of *Ascophyllum nodosum* (L.) Le Jolis. J. Exp. Mar. Biol. Ecol. 92:231–249.

Creese, R.G. 1988. Ecology of molluscan grazers and their interactions with marine algae in north-eastern New Zealand: a review. N.Z. J. Mar. Freshwater Res. 22:427–444.

Dakin, W.J. 1980. Australian seashores. Sydney: Angus and Robertson.

Dayton, P.K. 1975. Experimental evaluation of ecological dominance in a rocky intertidal community. Ecol. Monogr. 45:137–159.

Dayton, P.K. 1985. Ecology of kelp communities. Annu. Rev. Ecol. Syst. 16:215–245.

Dayton, P.K., Currie, V., Gerrodette, T., Keller, B.D., Rosenthal, R., and Ven Tresca, D. 1984. Patch dynamics and stability of southern California kelp communities. Ecol. Monogr. 54:253–289.

Denton, A., Chapman, A.R.O., and Markham, J. 1990. Size-specific concentrations of phlorotannins (antiherbivore compounds) in three species of *Fucus*. Mar. Ecol. Prog. Ser. 65:103–104.

de Nys, R., Coll, J.C., and Price, I.R. 1991. Chemically mediated interactions between the red alga *Plocamium hamatum* (Rhodophyta) and the octocoral *Sinularia cruciata* (Alcyonacea). Mar. Biol. 108:315–320.

DeWreede, R.E. 1979. Phenology of *Sargassum muticum* (Phaeophyta) in the Strait of Georgia, British Columbia. Syesis 11:1–9.

DeWreede, R.E. 1984. Growth and age class distribution of *Pterygophora californica* (Phaeophyta). Mar. Ecol. Prog. Ser. 19:93–100.

Druehl, L.D. 1980. The distribution of Laminariales in the North Pacific with reference to environmental influences. *In* Evolution today, ed. G.G.E. Scudder and J.L. Reveal, pp. 55–67. Proc. Second Int. Congr. Syst. Evol. Biol.

Dunlap, W.C., Chalker, B.E., and Oliver, J.K. 1986. Bathymetric adaptations of reef-building corals at Davies Reef, Great Barrier Reef, Australia. III. UV-B absorbing compounds. J. Exp. Mar. Biol. Ecol. 104:239–248.

Edmonds, J.S., and Francesconi, K.A. 1981a. Arseno-sugars from brown kelp (*Ecklonia radiata*) as intermediates in cycling of arsenic in a marine ecosystem. Nature 289:602–604.

Edmonds, J.S., and Francesconi, K.A. 1981b. The origin and chemical form of arsenic in the school whiting. Mar. Pollut. Bull. 12:92–96.

Estes, J.A., and Steinberg, P.D. 1988. Predation, herbivory and kelp evolution. Paleobiology 14:19–36.

Fain, S.R., Druehl, L.D., and Baillie, D.C. 1988. Repeat and single copy sequences are differentially conserved in the evolution of kelp chloroplast DNA. J. Phycol. 24:292–302.

Faulkner, D.J. 1984. Marine natural products: metabolites of marine algae and herbivorous marine molluscs. Nat. Prod. Rep. 1:251–280.

Faulkner, D.J. 1986. Marine natural products. Nat. Prod. Rep. 3:1–33.

Faulkner, D.J. 1987. Marine natural products. Nat. Prod. Rep. 4:539–576.

Feeny, P.P. 1968. Inhibitory effect of oak leaf tannins on larval growth of the winter moth *Operophtera brumata*. J. Insect Physiol. 14:805–817.

Feeny, P.P. 1969. Inhibitory effect of oak leaf tannins on the hydrolysis of proteins by trypsin. Phytochemistry 8:2119–2126.

Feeny, P.P. 1970. Seasonal changes in oak leaf tannins and nutrients as a cause of spring feeding by winter moth caterpillars. Ecology 51:565–581.

Feeny, P.P. 1976. Plant apparency and chemical defenses. Recent Adv. Phytochem. 10:1–42.

Fenical, W. 1980. Distributional and taxonomic features of toxin-producing marine algae. *In* Pacific seaweed aquaculture, ed. I.A. Abbott, M.S. Foster, and L.F. Eklund, pp. 144–151. California Sea Grant College Program. La Jolla: University of California.

Fenical, W. 1982. Diterpenoids. *In* Marine natural products, vol. 2, ed. P.J. Scheuer, pp. 173–245. New York: Academic Press.

Fletcher, W.J. 1987. Interactions among subtidal Australian sea urchins, gastropods and algae: effects of experimental removals. Ecol. Monogr. 57:89–109.

Fox, L.R. 1981. Defense and dynamics in plant-herbivore systems. Am. Zool. 21:853–864.

Freeland, W.J., Calcott, P.H., and Anderson, L.R. 1985a. Tannins and saponin: interaction in herbivore diets. Biochem. Syst. Ecol. 13:189–193.

Freeland, W.J., Calcott, P.H., and Geiss, D.P. 1985b. Allelochemicals, minerals, and herbivore population size. Biochem. Syst. Ecol. 13:195–206.

Gaines, S.D., and Lubchenco, J. 1982. A unified approach to marine plant-herbivore interactions. II. Biogeography. Annu. Rev. Ecol. Syst. 13:111–138.

Geiselman, J.A. 1980. Ecology of chemical defenses of algae against the herbivorous snail *Littorina littorea*, in the New England rocky intertidal community. Ph.D. dissertation, Massachusetts Institute of Technology, Cambridge, Mass.

Geiselman, J.A., and McConnell, O.J. 1981. Polyphenols in the brown algae *Fucus vesiculosus* and *Ascophyllum nodosum*: chemical defenses against the herbivorous snail *Littorina littorea*. J. Chem. Ecol. 7:1115–1133.

Gerwick, W.H., and Fenical, W. 1982. Phenolic lipids from related marine algae of the Order Dictyotales. Phytochemistry 21:633–637.

Gerwick, W.H., Fenical, W., and Norris, J.N. 1985. Chemical variation in the tropical seaweed *Stypopodium zonale* (Dictyotaceae). Phytochemistry 24:1279–1283.

Green, G. 1977. Ecology of toxicity in marine sponges. Mar. Biol. 40:207–215.

Gregson, R.P., Kazlauskas, R., Murphy, P.T., and Wells, R.J. 1982. New metabolites from the brown alga *Cystophora torulosa*. Austr. J. Chem. 30:2527–2532.

Grisebach, H. 1981. Lignins. *In* The biochemistry of plants. A comprehensive treatise, vol. 7., ed. E.E. Conn, pp. 457–478. New York: Academic Press.

Gschwend, P.M., MacFarlane, J.K., and Newman, K.A. 1985. Volatile halogenated organic compounds released to seawater from temperate marine macroalgae. Science 227:1033–1035.

Hagerman, A.E., and L.G. Butler. 1989. Choosing appropriate methods and standards for assaying tannins. J. Chem. Ecol. 15:1795–1810.

Hagerman, A.E., and Robbins, C.T. 1987. Implications of soluble tannin-protein complexes for tannin analysis and plant defense mechanisms. J. Chem. Ecol. 13:1243–1259.

Harrold, C., and Pearse, J.S. 1987. The ecological role of echinoderms in kelp forests. *In* Echinoderm studies, vol. 2, ed. M. Jangoux and J.M. Lawrence, pp. 137–233. Rotterdam: A.A. Balkema.

Harrold, C., and Reed, D.C. 1985. Food availability, sea urchin grazing, and kelp forest community structure. Ecology 66:1160–1169.

Haslam, E. 1981. Vegetable tannins. *In* The biochemistry of plants. A comprehensive treatise, vol. 7., ed. E.E. Conn, pp. 527–556. New York: Academic Press.

Hawkins, S.J., and Hartnoll, R.G. 1983. Grazing of intertidal algae by marine invertebrates. Oceanogr. Mar. Biol. Annu. Rev. 21:195–282.

Hay, M.E. 1981. Herbivory, algal distribution, and the maintenance of between habitat diversity on a tropical fringing reef. Am. Nat. 118:520–540.

Hay, M.E. 1984. Predictable spatial escapes from herbivory: how do these affect the evolution of herbivore resistance in tropical marine communities? Oecologia 64: 396–407.

Hay, M.E., Colburn, T., and Downing, D. 1983. Spatial and temporal patterns in herbivory on a Caribbean fringing reef: the effects on plant distribution. Oecologia 58:299–308.

Hay, M.E., Duffy, J.E., and Fenical, W. 1988c. Seaweed chemical defenses: among-compound and among-herbivore variance. *In* Proc. Sixth Int. Coral Reef Symp., vol. 3, pp. 43–48.

Hay, M.E., Duffy, J.E., and Fenical, W. 1990. Host-plant specialization decreases predation on a marine amphipod: an herbivore in plant's clothing. Ecology 71: 733–743.

Hay, M.E., Duffy, J.E., Fenical, W., and Gustafson, K. 1988b. Chemical defense in the seaweed *Dictyopteris delicatula*: differential effects against reef fishes and amphipods. Mar. Ecol. Prog. Ser. 48:185–192.

Hay, M.E., Duffy, J.E., Pfister, C.A., and Fenical, W. 1987b. Chemical defense against different marine herbivores: are amphipods insect equivalents? Ecology 68:1567–1580.

Hay, M.E., and Fenical, W. 1988. Marine plant-herbivore interactions: the ecology of chemical defense. Annu. Rev. Syst. Ecol. 19:111–145.

Hay, M.E., Fenical, W., and Gustafson, K. 1987a. Chemical diverse coral-reef herbivores. Ecology 68:1581–1592.

Hay, M.E., Renaud, P.E., and Fenical, W. 1988a. Large mobile vs. small sedentary herbivores and their resistance to seaweed chemical defenses. Oecologia 75:246–252.

Horn, M.H., Neighbors, M.A., Rosenberg, M.J., and Murray, S.N. 1985. Assimilation of carbon from dietary and nondietary macroalgae by a temperate-zone intertidal fish, *Cebidichthys violaceus* (Girard) (Teleostei: Stichaidae). J. Exp. Mar. Biol. Ecol. 86:241–253.

Ilvessalo, H., and Tuomi, J. 1989. Nutrient availability and accumulation of phenolic compounds in the brown alga *Fucus vesiculosus*. Mar. Biol. 101:115–119.

Janzen, D.H. 1979. New horizons in the biology of plant defenses. *In* Herbivores, ed. G.A. Rosenthal and D.H. Janzen, pp. 331–350. New York: Academic Press.

Johnson, C.R., and Mann, K.H. 1986. The importance of plant defense abilities to the structure of seaweed communities: the kelp *Laminaria longicruis* de la Pylaie survives grazing by the snail *Lacuna vincta* (Montagu) at high population densities. J. Exp. Mar. Biol. Ecol. 97: 231–267.

Jones, G.P., and Andrew, N.L. 1990. Herbivory and patch dynamics on rocky reefs in temperate Australasia: the roles of fish and sea urchins. Austr. J. Ecol. 15:505–520.

Katayama, T. 1951. Tannins of seaweeds. J. Chem. Soc. Japan Ind. Chem. Sect. 54:603–604.

Kato, T., Kumanireng, A.S., Ichinose, I., Kitihara, Y., Kakinada, Y., and Kato, Y. 1975. Structure and synthesis of active component from a marine alga, *Sargassum tortile*, which induces the settling of swimming larvae of *Coryne uchidai*. Chem. Lett., pp. 335–338.

Kazlauskas, R., King, L., Murphy, P.T., Warren, R.G., and Wells, R.J. 1981. New metabolites from the brown algal genus *Cystophora*. Aust. J. Chem. 34:439–447.

Keats, D.W., Steele, D.H., and South, G.R. 1984. Depth dependent reproductive output in the green sea urchin *Strongylocentrotus drobachiensis* (O.F. Miller) in relation to the nature and availability of food. J. Exp. Mar. Biol. Ecol. 80:77–82.

Kirkman, H. 1984. Standing stock and production of *Ecklonia radiata* (C.Ag.) J. Agardh. J. Exp. Mar. Biol. Ecol. 76:119–130.

Klinger, T., and DeWreede, R.E. 1988. Stipe rings, age and size in populations of *Laminaria setchellii* Silva (Laminariales, Phaeophyta) in British Columbia, Canada. Phycologia 27:234–40.

Koch, M., and Glombitza, K.W. 1980. Phlorotannins from *Laminaria ochroleuca*. Phytochemistry 19:1821–1823.

Kurata, K., Taniguchi, K., Shiraishi, K., Hayama, N., Tanaka, I., and Suzuki, M. 1989. Ecklonialactone-A and -B, two unusual metabolites from the brown alga *Ecklonia stolonifera* Okamura. Chem. Lett., pp. 267–270.

Larson, B.R., Vadas, R.L., and Keser, M. 1980. Feeding and nutritional ecology of the sea urchin *Strongylocentrotus drobachiensis* in Maine, USA. Mar. Biol. 59:49–62.

Lawrence, J. 1975. On the relationship between marine plants and sea urchins. Oceanogr. Mar. Biol. Annu. Rev. 13:213–246.

Lewis, J.R. 1964. The ecology of rocky shores. London: English Universities Press.

Littler, M.M., Taylor P.R., and Littler, D.S. 1983. Algal resistance to herbivory on a Caribbean barrier reef. Coral Reefs 2:111–118.

Littler, M.M., Taylor, P.R., and Littler, D.S. 1986. Plant defense associations in the marine environment. Coral Reefs 5:63–72.

Lobban, C.S., Harrison, P.J., and Duncan, M.J. 1985. The physiological ecology of seaweeds. Cambridge: Cambridge University Press.

Lowe, E.F., and Lawrence, J.M. 1976. Absorption efficiencies of *Lytechinus variegatus* (Lamarck) (Echinodermata: Echinoidea) for selected marine plants. J. Exp. Mar. Biol. Ecol. 21:223–234.

Lubchenco, J. 1978. Plant species diversity in a marine intertidal community: importance of herbivore food preference and algal competitive abilities. Am. Nat. 112:23–39.

Lubchenco, J. 1982. Effects of grazers and algal competitors on fucoid colonization in tide pools. J. Phycol. 18:544–550.

Lubchenco, J. 1983. *Littorina* and *Fucus*: effects of herbivores, substratum heterogeneity, and plant escapes during succession. Ecology 64:1116–1123.

Lubchenco, J., and Gaines, S.D. 1981. A unified approach to marine plant-herbivore interactions. I. Populations and communities. Annu. Rev. Ecol. Syst. 12:405–437.

Macauley, B.J., and Fox, L.R. 1980. Variation in total phenols and condensed tannins in *Eucalyptus*: leaf phenology and insect grazing. Aust. J. Ecol. 5:31–35.

McConnell, O.J., Hughes, P.A., Targett, N.M., and Daley, J. 1982. Effects of secondary metabolites on feeding by the sea urchin, *Lytechinus variegatus*. J. Chem. Ecol. 8:1427–1453.

McEnroe, F.J., Robertson, K.J., and Fenical, W. 1977. Diterpenoid synthesis in brown seaweeds in the family Dictyotaceae. *In* Marine natural products chemistry, ed. D.J. Faulkner and W. Fenical, pp. 179–189. New York: Plenum Press.

McKey, D. 1979. Distribution of secondary compounds in plants. *In* Herbivores, ed. G.A. Rosenthal and D.H. Janzen, pp. 56–133. New York: Academic Press.

McLure, J.W. 1979. The physiology of phenolic compounds in plants. Recent Adv. Phytochem. 12:525–555.

Mann, K.H. 1972. Ecological energetics of the sea-weed zone in a marine bay on the Atlantic coast of Canada. II. Productivity of the seaweeds. Mar. Biol. 14:199–209.

Mann, K.H. 1973. Seaweeds: their productivity and strategy of growth. Science 182:975–981.

Mann, K.H. 1982. Ecology of coastal waters. Oxford: Blackwell Scientific.

Martin, J.S., and Martin, M.M. 1982. Tannin assays in ecological studies. J. Chem. Ecol. 9:285–294.

Martin, J.S., Martin, M.M., and Bernays, E.A. 1987. Failure of tannic acid to inhibit digestion or reduce digestibility of plant protein in gut fluids of insect herbivores: implications for theories of plant defense. J. Chem. Ecol. 13:605–621.

Martin, M.M., and Martin, J.S. 1984. Surfactants: their role in preventing the precipitation of proteins by tannins in insect guts. Oecologia 61:342–345.

Mayer, A.M., and Harel, E. 1979. Polyphenol oxidases in plants. Phytochemistry 18:193–215.

Mehansho, H., Butler, L.G., and Carlson, D.M. 1987. Dietary tannins and salivary proline-rich proteins: interactions, induction, and defense mechanisms. Annu. Rev. Nutr. 7:423–440.

Mole, S., and Waterman, P.G. 1987. A critical analysis of techniques for measuring tannins in ecological studies. I. Techniques for chemically defining tannins. Oecologia 72:137–147.

Moss, B.L. 1983. Sieve elements in the Fucales. New Phytol. 93:433–437.

Nizamuddin, M. 1970. Phytogeography of the Fucales and their seasonal growth. Bot. Mar. 13:131–139.

Norton, T.A. 1977. The growth and development of *Sargassum muticum* (Yendo) Fensholt. J. Exp. Mar. Biol. Ecol. 26:41–53.

Novaczek, I. 1984. Development and phenology of *Ecklonia radiata* at two depths in Goat Island Bay, New Zealand. Mar. Biol. 81:189–197.

Padilla, D.K. 1985. Structural resistance of algae to herbivores. Mar. Biol. 90:103–109.

Pathirana, C., and Anderson, R.J. 1984. Diterpenoids from the brown alga *Dictyota binghamiae*. Can. J. Chem. 62:1666–1670.

Paul, V.J., and Fenical, W. 1986. Chemical defense in tropical green algae, order Caulerpales. Mar. Ecol. Prog. Ser. 34:157–169.

Paul, V.J., Hay, M.E., Duffy, J.E., Fenical, W., and Gustafson, K. 1987. Chemical defense in the seaweed *Ochtodes secundiramea* (Montagne) Howe (Rhodophyta): effects of its monoterpenoid components upon diverse coral-reef herbivores. J. Exp. Mar. Biol. Ecol. 114:249–260.

Paul, V.J., and Van Alstyne, K.L. 1988. Chemical defense and chemical variation in some tropical Pacific species of *Halimeda* (Halimedaceae: Chlorophyta). Coral Reefs 6:243–264.

Pederson, A. 1984. Studies on phenol content and heavy metal uptake in fucoids. *In* Proc. Int. Seaweed Symp., vol. 11, pp. 498–504.

Pielou, E.C. 1977. The latitudinal spans of seaweed species and their patterns of overlap. J. Biogeogr. 4:299–311.

Quast, J.C. 1968. Observations on the food of kelp bed fishes. Bull. Calif. Fish. Game 139:109–142.

Ragan, M.A. 1976. Physodes and the phenolic structures of brown algae. Composition and the structure of physodes in vivo. Bot. Mar. 19:145–154.

Ragan, M.A. 1985. The high-molecular-weight polyphloroglucinols of the marine brown alga *Fucus vesiculosus* L.: degradative analysis. Can. J. Chem. 63:294–303.

Ragan, M.A., and Craigie, J.S. 1980. Quantitative studies on brown algal phenols. IV. Ultraviolet spectrophotometry of extracted polyphenols and implications for measuring dissolved organic matter in seawater. J. Exp. Mar. Biol. Ecol. 46:231–239.

Ragan, M.A., and Glombitza, K.W. 1986. Phlorotannins, brown algal polyphenols. Prog. Phycol. Res. 4:129–241.

Ragan, M.A., and Jensen, A. 1977. Quantitative studies on brown algal polyphenols. I. Estimation of absolute polyphenol content of *Ascophyllum nodosum* (L.) and *Fucus vesiculosus* (L.). J. Exp. Mar. Biol. Ecol. 34:245–258.

Ragan, M.A., and Jensen, A. 1978. Quantitative studies on brown algal phenols. II. Seasonal variation in polyphenol content of *Ascophyllum nodosum* (L.) Le Jol. and *Fucus vesiculosus* (L.). J. Exp. Mar. Biol. Ecol. 34:245–258.

Rhoades, D. 1979. Evolution of plant chemical defenses against herbivores. *In* Herbivores, ed. G.A. Rosenthal and D.H. Janzen, pp. 4–54. New York: Academic Press.

Rhoades, D.F., and Cates, R.G. 1976. Toward a general theory of plant antiherbivore chemistry. Recent Adv. Phytochem. 10:168–213.

Ridley, M. 1983. The explanation of organic diversity: the comparative method and adaptations for mating. Oxford: Oxford University Press.

Robb, M.F. 1975. The diet of the chiton *Cyanoplax hartwegii* in three intertidal habitats. Veliger 18 (Suppl.):34–37.

Robbins, C.T., Hanley, T.A., Hagerman, A.E., Hjeljord, O., Baker, D.L., Schwartz, C.C., and Mautz, W.W. 1987. Role of tannins in defending plants against ruminants: reduction in protein availability. Ecology 68:98–107.

Ronneberg, O., and Ruokolahti, C. 1986. Seasonal variation of algal epiphytes and phenolic content of *Fucus vesiculosus* in a northern Baltic archipelago. Ann. Bot. Fenn. 23:317–323.

Russ, G. 1984a. Distribution and abundance of herbivorous grazing fishes in the central Great Barrier Reef. I. Levels of variability across the entire continental shelf. Mar. Ecol. Prog. Ser. 20:23–34.

Russ, G. 1984b. Distribution and abundance of herbivorous grazing fishes in the central Great Barrier Reef. II. Patterns of zonation of mid-shelf and outershelf reefs. Mar. Ecol. Prog. Ser. 20:35–44.

Santelices, B., Castilla, J.C., Cancino, J., and Schmiede, P. 1980. The comparative ecology of *Lessonia nigrescens* and *Durvillea antarctica* in central Chile. Mar. Biol. 59:119–132.

Santelices, B., and Ojeda, F.P. 1984. Effects of canopy removal on the understory algal community structure of coastal forests of *Macrocystis pyrifera* from southern South America. Mar. Ecol. Prog. Ser. 14:165–173.

Schiel, D.R. 1982. Selective feeding by the echinoid, *Evechinus chloroticus*, and the removal of plants from subtidal algal stands in northern New Zealand. Oecologia 54:379–388.

Schiel, D.R. 1985. A short-term demographic study of *Cystoseira osmundacea* (Fucales: Cystoseiraceae) in central California. J. Phycol. 21:99–106.

Schiel, D.R., and Foster, M.S. 1986. The structure of subtidal algal stands in temperate waters. Oceanogr. Mar. Biol. Annu. Rev. 24:265–308.

Schmitz, K., and Lobban, C.S. 1976. A survey of translocation in Laminariales (Phaeophyceae). Mar. Biol. 36:207–216.

Schonbeck, M.W., and Norton, T.A. 1978. Factors controlling the upper limits of fucoid algae on the shore. J. Exp. Mar. Biol. Ecol. 31:303–313.

Schultz, J.C., and Baldwin, I.T. 1982. Oak leaf quality declines in response to defoliation by gypsy moth larvae. Science 217:149–151.

Scott, F.J., and Russ, G.R. 1987. Effects of grazing on species composition of the epilithic algal community on coral reefs of the central Great Barrier Reef. Mar. Ecol. Prog. Ser. 39:293–304.

Shizuru, Y., Matsukawa, S., Ojika, M., and Yamada, K. 1982. Two new farnesylacetone derivatives from the brown alga *Sargassum micracanthum*. Phytochemistry 21:1808–1809.

Sieburth, J.M. 1968. The influence of algal antibiosis on the ecology of marine microorganisms. *In* Advances in microbiology of the sea, ed. M.R. Droop and E.F.J. Wood, pp. 63–94. New York: Academic Press.

Sieburth, J.M., and Conover, J.T. 1965. *Sargassum* tannin, an antibiotic which retards fouling. Nature 208:52–53.

Sims, E.L., and Rausher, M.D. 1987. Costs and benefits of plant resistance to herbivory. Am. Nat. 130:570–581.

Smith, R.C., and and Baker, K.S. 1979. Penetration of UV-B and biologically effective dose rates in natural waters. Photochem. Photobiol. 29:311–323.

Sousa, W.P. 1979. Experimental investigations of disturbance and ecological succession in a rocky intertidal algal community. Ecol. Monogr. 49:227–254.

Steinberg, P.D. 1984. Algal chemical defense against herbivores; allocation of phenolic compounds in the kelp *Alaria marginata*. Science 223:405–407.

Steinberg, P.D. 1985. Feeding preferences of *Tegula funebralis* and chemical defenses in marine brown algae. Ecol. Monogr. 55:333–349.

Steinberg, P.D. 1986. Chemical defenses and the susceptibility of tropical marine algae to herbivores. Oecologia 69:628–630.

Steinberg, P.D. 1988. The effects of quantitative and qualitative variation in phenolic compounds on feeding in three species of marine invertebrate herbivores. J. Exp. Mar. Biol. Ecol. 120:221–237.

Steinberg, P.D. 1989. Biogeographical variation in brown algal polyphenolics and other secondary metabolites: comparison between temperate Australasia and North America. Oecologia 78:374–383.

Steinberg, P.D., Edyvane, K., DeNys, R., Birdsey, R., and van Altena, I.A. 1991. Lack of avoidance of phenolic-rich algae by tropical herbivorous fishes. Mar. Biol. 109:335–344.

Steinberg, P.D., and Paul, V.J. 1990. Fish feeding and chemical defenses of tropical brown algae in Western Australia. Mar. Ecol. Prog. Ser. 58:253–259.

Steinberg, P.D., and van Altena, I.A. 1992. Chemical defenses in temperate Australasian seaweeds. I. Tolerance of invertebrate herbivores to algal polyphenolics. Ecol. Monogr., in press.

Steneck, R.S. 1983. Escalating herbivory and resulting adaptive trends in calcareous algal crusts. Paleobiology 9:44–61.

Steneck, R.S. 1986. The ecology of coralline algal crusts: convergent patterns and adaptive strategies. Annu. Rev. Ecol. Syst. 17:273–303.

Steneck, R.S., and Watling, L. 1982. Feeding capabilities and limitations of herbivorous molluscs: a functional group approach. Mar. Biol. 68:299–319.

Swain, T. 1977. Secondary compounds as protective agents. Annu. Rev. Plant Physiol. 28:470–501.

Swain, T. 1978. Plant-animal coevolution: a synoptic view of the paleozoic and mesozoic. In Biochemical aspects of plant and animal coevolution, ed. J.B. Harborne, pp. 309–323. New York: Academic Press.

Swain, T. 1979. Tannins and lignins. In Herbivores, ed. G.A. Rosenthal, and D.H. Janzen, pp. 657–682. New York: Academic Press.

Swain, T., and Hillis, W.E. 1959. The phenolic constituents of Prunus domesticus. I. The quantitative analysis of phenolic constituents. J. Sci. Food Agric. 10:63–68.

Targett, N.M., Coen, L.D., Boettcher, A.A., and Tanner, C.E. Submitted. Biogeographic comparisons of marine algal polyphenolics: evidence against a latitudinal trend. Oecologia.

Tempel, A.S. 1982. Tannin-measuring techniques: a review. J. Chem. Ecol. 8:1289–1298.

Tremblay, C., and Chapman, A.R.O. 1980. The local occurrence of Agarum cribrosum in relation to the presence or absence of its competitors and predators. Proc. Nova Scotia Inst. Sci. 30:165–170.

Tugwell, S., and Branch, G.A. 1989. Differential phenolic distribution among tissues in the kelps Ecklonia maxima, Laminaria pallida, and Macrocystis angustifolia in relation to plant defense theory. J. Exp. Mar. Biol. Ecol. 129:219–230.

Underwood, A.J. 1980. The effects of grazing by gastropods and physical factors on the upper limits of distribution of intertidal macroalgae. Oecologia 46:201–213.

Underwood, A.J., and Jernakoff, P. 1981. Effects of interactions between algae and grazing gastropods on the structure of a low shore intertidal algal community. Oecologia 48:221–223.

Vadas, R.L. 1977. Preferential feeding: an optimization strategy in sea urchins. Ecol. Monogr. 47:337–371.

Van Alstyne, K.L. 1988. Herbivore grazing increases polyphenolic defenses in the intertidal brown alga Fucus distichus. Ecology 69:655–663.

Van Alstyne, K.L., and Paul, V.J. 1988. The role of secondary metabolites in marine ecological interactions. In Proc. Sixth Int. Coral Reef Symp., vol. 1, pp. 175–186.

Van Alstyne, K.L., and Paul, V.J. 1990. The evolution and biogeography of antiherbivore compounds in marine macroalgae: why don't tropical brown algae use temperate defenses against herbivorous fishes? Oecologia 84:158–163.

van Altena, I.A. 1988. Terpenoids from the brown alga Cystophora moniliformis. Aust. J. Chem. 41:49–56.

Vermeij, G.J. 1978. Biogeography and adaptation. Cambridge: Harvard University Press.

Watson, D.C., and Norton, T.A. 1985. Dietary preferences of the common periwinkle, Littorina littorea (L.). J. Exp. Mar. Biol. Ecol. 88:193–211.

Womersley, H.B.S. 1967. A critical survey of the marine algae of South Australia. II. Phaeophyta. Aust. J. Bot. 15:189–270.

Womersley, H.B.S. 1981. Biogeography of Australian marine macroalgae. In Marine botany, an Australasian perspective, pp. 292–307. Ayreshire: Longman.

Wylie, C.R., and Paul, V.J. 1988. Feeding preferences of the surgeonfish *Zebrasoma flavescens* in relation to chemical defenses of tropical algae. Mar. Ecol. Prog. Ser. 45:23–32.

Wynne, M.J., and Loiseaux, S. 1976. Recent advances in life history studies of the Phaeophyta. Phycologia 15:435–452.

Zucker, W. 1983. Tannins: does structure determine function? An ecological perspective. Am. Nat. 121:335–365.

Chapter 3
The Role of Seaweed Chemical Defenses in the Evolution of Feeding Specialization and in the Mediation of Complex Interactions

MARK E. HAY

Herbivory on seaweeds is intense and is often the primary factor affecting their distribution and abundance (Lubchenco and Gaines 1981; Hay 1985; Lewis 1986; Schiel and Foster 1986). This is especially true on tropical coral reefs, where either sea urchins or fishes alone can consume nearly 100% of local production (Carpenter 1986). In reef systems as well as in temperate habitats such as kelp beds, grazers appear to have been a primary agent selecting for seaweeds with both chemical and morphological characteristics that deter herbivores (Steneck 1983; Estes and Steinberg 1988). Several recent reviews address the direct effects of seaweed chemical defenses (Paul and Fenical 1987; Hay and Fenical 1988; Hay et al. 1988d; Duffy and Hay 1990; Hay 1991a, 1991b; Hay and Steinberg 1992; Paul, this volume). In this chapter I address several less obvious, more complex, and more indirect interactions that may be affected by seaweed chemical defenses. These include (1) the role of seaweed metabolites in the evolution of feeding specialization, (2) the chemical mediation of seaweed associational resistance to herbivores, (3) the potential interactions among different metabolites produced by individual seaweeds, and (4) other indirect effects of seaweed metabolites such as their potential effects on the rate of carbon and nutrient cycling.

Generalist versus Specialist Herbivores and the Evolution of Seaweed Chemical Defenses

Marine communities differ dramatically from terrestrial communities in the impact of specialist herbivores on plant populations and communities. Insect herbivory is very important in terrestrial communities, and the majority of insect herbivores are relatively specialized (Futuyma and Gould 1979; Price 1983; Strong et al. 1984). As one example, more than 80% of all North American butterflies feed only on members of one plant family (Ehrlich and Murphy 1988). In marine communities, generalist herbivores such as fishes, sea urchins, and some gastropods are the dominant herbivores (Lubchenco and Gaines 1981; Hawkins and Hartnoll 1983; Hay 1985; Carpenter 1986; Lewis 1986; Schiel and Foster 1986; Morrison 1988), and the majority of these species are extremely generalized feeders. As an example, Figure 3.1a shows a conservative assessment of the number of plant families eaten by herbivorous fishes in the Caribbean. In contrast to the butterfly example mentioned above, the average herbivorous fish species feeds on more than 7.4 families and 3.8 divisions of marine plants. Sea urchins may be even more generalized feeders (Littler et al. 1983; Morrison 1988). The Japanese urchin *Strongylocentrotus intermedius* and the Caribbean urchin *Diadema antillarum* appear to be representative; both feed on more than 20 families of seaweeds (Ogden et al. 1973; Kawamura 1973). Lowe's (1975) study of the Caribbean urchin *Lytechinus variegatus* provides an additional example. After reporting the numerous seaweeds and animals consumed by this urchin, he noted that it would also eat aquarium air stones, cork, filter paper, rubber stoppers, and swiss cheese.

Selective pressures for the evolution of characteristics that deter herbivores are therefore usually generated by a diverse assemblage of generalist herbivores with differing mobilities, habitat requirements, feeding modes, and digestive physiologies. In the face of this type of diffuse herbivore pressure (Fox 1981), selection should favor seaweed traits that are broadly active against numerous types of herbivores. Since single groups of herbivores such as fishes alone or urchins alone are capable of consuming all algal production in some habitats (Carpenter 1986), the evolution of species-specific or even group-specific herbivore deterrents may be of limited value unless they are coupled with additional defenses that deter other herbivores as well. This may

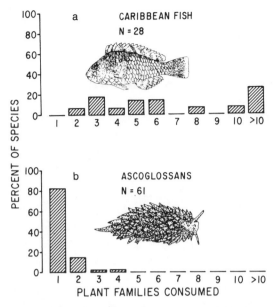

Figure 3.1 (a) The number of plant families known to be consumed by each species of Caribbean fish listed by Randall (1967) as having gut contents in which more than 75% of the volume was plant material. These data are conservative because they were derived from gut contents of a limited number of specimens for each species. Additionally, all species of cyanobacteria (blue-green algae) were arbitrarily counted as coming from one family, as were all diatoms. This prevented a bias in the samples from a large number of families that could have made up a minimum of the gut volume and might have been eaten primarily as epiphytes on other foods. (b) The number of plant families consumed by the herbivorous ascoglossans listed by Jensen (1980).

be why seaweeds on herbivore-rich coral reefs so commonly employ combinations of structural, morphological, and chemical defenses and, in some cases, coordinate these defenses with patterns of temporal and microhabitat escape (Hay 1984; Paul and Hay 1986; Lewis et al. 1987; Hay and Fenical 1988; Hay et al. 1988b; Paul and Van Alstyne 1988a).

In contrast to terrestrial systems, specialist marine herbivores, such as the ascoglossan gastropods as a group (see Fig. 3.1b; Jensen 1980) or a few unusual specialist amphipods (Hay et al. 1990a) and crabs (Hay et al. 1989, 1990b), appear to have minimal effects on their host plants (Clark and DeFreeze 1987; Hay et al. 1989). Thus, specialist marine herbivores probably exert little selective pressure for the evolution of defensive metabolites relative to the extreme pressure exerted by generalists.

Why Small Herbivores Eat Noxious Plants

Numerous previous studies have demonstrated that many seaweed metabolites significantly deter feeding by mixed species groups of reef fishes and by sea urchins; these compounds thus provide a selective advantage under natural conditions (see references in Hay and Fenical 1988). But secondary metabolites that deter feeding by generalist fishes and urchins often fail to deter, and may even stimulate, feeding by generalist or specialist herbivores that are small, less mobile, and subject to high potential rates of predation (Hay et al. 1987a, 1988a, 1988c, 1989, 1990a, 1990b; Paul et al. 1987; Paul and Van Alstyne 1988b; Duffy and Hay 1991a). Hay and his co-workers have argued that once seaweeds evolve effective defenses against fishes, these algae then become microsites of relative safety for small grazers of limited mobility that live in close association with the plants they consume. These small sedentary herbivores (e.g., some amphipods, crabs, and polychaetes—often called mesograzers) can be severely affected by predators (Vince et al. 1976; Van Dolah 1978; Young et al. 1976; Young and Young 1978; Nelson 1979a, 1979b; Stoner 1979, 1980; Edgar 1983a, 1983b). If the ability to live on and feed from seaweeds that are avoided by fishes significantly lowers their susceptibility to predation, then these animals may be under strong selection to circumvent seaweed chemical defenses. Similar arguments have been made for the importance of "enemy-free space" in the evolution of habitat and feeding specialization among terrestrial insect herbivores (Gilbert and Singer 1975; Lawton 1978; Price et al. 1980, 1986; Bernays and Graham 1988; Bernays 1989; Bernays and Cornelius 1989). In marine communities, however, it appears that generalist mesograzers of limited mobility may have evolved resistance to a broad range of unrelated lipophilic metabolites, and although these mesograzers preferentially feed on seaweeds repellent to fishes, they do not often become restrictively specialized to certain host plants (Hay et al. 1988a, 1988c; Duffy and Hay 1991a) as is common in terrestrial communities.

Generalist Mesograzers and the Importance of Mobility, Size, and Predator Escape

Large mobile herbivores like fishes may move among and feed from thousands of plants each day. Since they do not commonly live in intimate association with their food plants for extended periods, they utilize seaweeds as foods rather than living sites. For smaller herbivores that spend significant periods living on the algae they consume, seaweeds often function as both food and habitat. These differences between large mobile versus small sedentary herbivores may significantly affect the selective regime operating on each type of herbivore even if both are generalist feeders (Hay et al. 1988a, 1988c; Duffy and Hay 1990, 1991a, 1991b; Hay 1991b).

Studies in terrestrial, freshwater, and marine habitats have shown that risk of predation significantly affects how consumers forage, and that foragers may exchange food return for greater safety when the two factors conflict (Sih 1980; Hay and Fuller 1981; Werner et al. 1983; Power 1984; Hay 1985; Damman 1987; Holbrook and Schmitt 1988). Since fish predation on marine mesograzers is intense, mesograzers that associate with individual plants may lower their rate of encounter with omnivorous and herbivorous fishes by living on seaweeds that are not eaten, and thus seldom visited, by these fishes.

A recent experimental study (Duffy 1989) on two species of amphipods that differ considerably in their mobilities and in the length of time they spend on individual seaweeds suggests that less mobile mesograzers can decrease predation by associating with chemically defended seaweeds. *Ampithoe longimana* builds mucilaginous tubes on seaweeds and feeds near the mouth of these tubes. In laboratory experiments, *A. longimana* spent lengthy periods of time on individual plants and experienced less predation when on the chemically defended alga *Dictyota menstrualis* than when on the palatable (to fishes) alga *Ulva*. Feeding by this amphipod is stimulated or unaffected by *Dictyota*'s secondary metabolites, pachydictyol-A (1) and dictyol-E (2),

1   2

which significantly deter feeding by co-occurring fishes (Hay et al. 1987a, 1988c). In contrast, the highly mobile amphipod *Gammarus mucronatus* does not build tubes, often moves among different plants, is not resistant to *Dictyota*'s chemical defenses, and does not experience lower rates of fish predation when associated with *Dictyota* versus *Ulva*. During periods of maximal fish abundance, *G. mucronatus* was rarely found in association with *Dictyota* and its population decreased to near local extinction. During this same period, numbers of *A. longimana* increased dramatically and they were found primarily in association with *Dictyota*. Because *A. longimana* does not sequester deterrent metabolites from *Dictyota* (Hay et al. 1987a), its ability to increase in numbers during periods of increased predator activity appears to be related primarily to its close physical association with *Dictyota* (Duffy and Hay 1991b).

Although this is the first study to assess how seaweed-mesograzer-predator interactions vary among closely related mesograzers that differ in mobility, several earlier studies noted that mesograzers often selectively consumed seaweeds that were avoided by fishes and that metabolites from these seaweeds significantly deterred feeding by the fishes but stimulated or did not affect feeding by the mesograzers. As examples, in coastal North Carolina, the omnivorous fishes *Lagodon rhomboides* and *Diplodus holbrooki* were deterred from feeding by the diterpene alcohols pachydictyol-A and dictyol-E, which occur in *Dictyota menstrualis*; the tubiculous amphipod *Ampithoe longimana* and the tubiculous polychaete *Platynereis dumerilii* were unaffected or stimulated to feed by these compounds, and they preferentially consumed *Dictyota* (Hay et al. 1987a, 1988c).

Studies of different species in the Caribbean showed a similar pattern (Hay et al. 1988a). The amphipod *Hyale macrodactyla* was abundant on the brown seaweed *Dictyopteris delicatula*, which it selectively consumed. This seaweed produced the C_{11} hydrocarbons dictyopterene-A (3) and dictyopterene-B (4),

3   4

which significantly deterred feeding by reef fishes but had no effect on feeding by the amphipods (Hay et al. 1988a). Feeding by these amphipods was also unaffected by the monoterpenoid ochtodene (5)

5

from the red alga *Ochtodes secundiramea*, even though this compound strongly deterred reef fishes (Paul et al. 1987). Although mesograzers in the above studies were stimulated or unaffected by compounds that deterred fishes, and although they may have experienced reduced predation by associating with these chemically defended seaweeds, none of the mesograzers was restrictively specialized to the chemically defended plants on which they preferentially fed (Hay et al. 1988a, 1988c).

Do Specialist Herbivores Escape or Deter Predators by Associating with Noxious Host Plants?

Both field experimentation and comparative descriptive studies indicate that mesograzer populations are often limited by predation and seldom limited by food (Vince et al. 1976; Van Dolah 1978; Nelson 1979a, 1979b; Stoner 1979, 1980; Edgar 1983a, 1983b). When unusual environmental circumstances free mesograzers from predation, some species may deplete their food resources (e.g., decimate entire kelp beds; Jones 1965; Tegner and Dayton 1987; Dayton and Tegner 1990); however, the rarity of such events suggests that mesograzer populations are usually kept well below carrying capacity by high rates of predation. Rigorous review of the literature on plant-insect interactions in terrestrial systems indicates that insects also are seldom food limited and often appear to be predator limited (Strong et al. 1984). It does not therefore seem reasonable to propose resource partitioning as the dominant factor driving the restrictively specialized feeding habits of many abundant groups of herbivorous terrestrial insects. This has generated considerable controversy regarding the evolutionary forces selecting for feeding specialization in insects; much of this controversy has centered on the relative importance of "enemy-free space" versus other factors in selecting for host plant specificity (Gilbert and Singer 1975; Lawton 1978; Price et al. 1980, 1986; Barbosa 1988; Bernays and Graham 1988; Courtney 1988; Ehrlich and Murphy 1988; Fox 1988; Janzen 1988; Jermy 1988; Rausher 1988; Schultz 1988; Thompson 1988).

Determining the importance of different factors in selecting for host plant specialization may be easier in marine than in terrestrial communities because there are many fewer specialists in marine communities. In addition, the rarity of specialist herbivores may mean that specialization in marine communities occurs only under a limited number of intense selective regimes. If this is the case, then common factors selecting for (or against) specialization may be more obvious in marine than in terrestrial communities. The limited data available on specialized marine herbivores are consistent with the hypothesis that predator avoidance and deterrence are major advantages arising from host plant specialization. Studies supporting this contention are outlined below.

Specialist Herbivores That Do Not Sequester Defensive Metabolites   Small herbivores could decrease their losses to predators if they were rarely encountered by predators or if they were not attacked when encountered. By specializing on chemically defended plants that are seldom visited by herbivorous fishes, sedentary mesograzers may decrease encounters with herbivores. This strategy could become even more effective if the mesograzers were cryptic when on their host plants. Three studies evaluated how specializing on chemically defended seaweeds affected predation on mesograzers that do not physiologically sequester defensive metabolites from their algal hosts; in each case specialization significantly reduced susceptibility to predation (Hay et al. 1989, 1990a, 1990b).

In the Caribbean, the herbivorous crab *Thersandrus compressus* is found only on the green seaweed *Avrainvillea longicaulis*, the only alga it eats in laboratory assays (Hay et al. 1990b). The crab's flattened morphology and the feltlike hairy protrusions from its legs and body make it nearly invisible when on the alga, which is also feltlike. *Avrainvillea* is a relatively low preference food for reef fishes (Hay 1984; Paul and Hay 1986), and it produces a brominated diphenylmethane derivative called avrainvilleol (6) (Sun et al. 1983)

6

that significantly deters fish feeding in field assays (Hay et al. 1990b). Although *Thersandrus* specializes on *Avrainvillea*, it does not sequester avrainvilleol. Crabs were immediately eaten when dropped into aquaria with the wrasse *Thalassoma bifasciatum*, a common predator on reef mesograzers. When crabs were dropped into these same tanks on pieces of *Avrainvillea* only slightly bigger than the crab itself, the crabs were not attacked, apparently because they were not recognized by the fish (Hay et al. 1990b).

On Australia's Great Barrier Reef, the crab *Caphyra rotundifrons* lives only in patches of the chemically defended green alga *Chlorodesmis fastigiata*. It eats only this alga and is impossible to see when hidden among the alga's filaments (Hay et al. 1989). *Chlorodesmis* produces the unique diterpenoid metabolite chlorodesmin (7) (Wells and Barrow 1979).

7

Both this compound and the crude organic extract of *Chlorodesmis* significantly deter feeding by Pacific reef fishes (Paul 1987; Wylie and Paul 1988). In contrast, chlorodesmin significantly stimulates feeding by the crab *Caphyra rotundifrons* (Hay et al. 1989). Although *Caphyra* restrictively specializes on *Chlorodesmis* and is stimulated to feed by the cytotoxic (Paul and Fenical 1987) compound that deters reef fishes, it does not sequester chlorodesmin. When crabs were tethered in different microhabitats on reefs, those tethered in *Chlorodesmis* were rarely eaten by fishes; those tethered in the open or in patches of the palatable (to fishes) green alga *Chaetomorpha* were all rapidly eaten. Thus, both the Australian and Caribbean crabs became relatively immune to predation by simple physical association with a host plant that was chemically resistant to fish grazing.

As a final example, the amphipod *Pseudamphithoides incurvaria* lives in a mobile, bivalved domicile that it constructs from the chemically defended brown alga *Dictyota bartayresii* (Lewis and Kensley 1982). Hay et al. (1990a) studied the chemically mediated interactions of this amphipod with its algal host and predators. Chemical assays demonstrated that natural populations of *Pseudamphithoides* constructed their domiciles from *Dictyota bartayresii* even when this alga was rarer than other *Dictyota* species and other related genera in the family Dictyotaceae. In both choice and no-choice tests in the laboratory, *Pseudamphithoides* built domiciles from and selectively consumed species of *Dictyota* that produced dictyol-class diterpenes that deterred feeding by reef fishes. Other brown seaweeds in the family Dictyotaceae—including a *Dictyota* species—that did not produce these deterrents to fish feeding were avoided by the amphipod. Amphipods removed from their domiciles were rapidly eaten when presented to predatory fish; amphipods in their domiciles were consistently rejected by fish. When amphipods were forced to build domiciles from the palatable (to fishes) alga *Ulva*, those in *Ulva* domiciles were eaten when offered to predatory wrasses while amphipods in *Dictyota* domiciles were rejected.

Algal defensive chemistry directly cued domicile building. When the green alga *Ulva* was treated with pachydictyol-A (1) (the major secondary metabolite produced by *Dictyota bartayresii*), domicile building by *Pseudamphithoides* increased in direct proportion to the concentration of pachydictyol-A. The chemical cuing of *Pseudamphithoides'* domicile-building behavior suggests that this could be considered "behavioral sequestering" of algal chemical defenses by an organism that cannot sequester metabolically. *Pseudamphithoides* differs from the crabs mentioned above in that it not only escapes predators by hiding in its algal domicile, but the domicile also deters predators that do recognize and attack the amphipod. All data collected during this study are consistent with the hypothesis that predator escape and deterrence are primary factors selecting for host specialization by *Pseudamphithoides incurvaria*.

Although all of the above are examples of chemical mediation, specialized mesograzers might be able to use a variety of algal characteristics other than chemistry to reduce their susceptibility to predation. As possible examples, some limpets feed almost exclusively on smooth coralline algae that provide an excellent attachment site and lower the limpet's susceptibility to predators that pry them from the substrate (Steneck 1982; R. S. Steneck pers. comm., 1991). Other limpets that specialize on kelps form grazing depressions on their thick algal hosts; these grazing scars fit tightly around their shells and substantially lower their risk of being dislodged by predatory starfish (Phillips and Castori 1982).

Herbivores That Deter Predation Using Sequestered Algal Metabolites   Opisthobranch gastropods in the order Ascoglossa are among the best studied of the truly specialized marine herbivores. They are the only marine herbivores that, as a group, show a high degree of feeding specialization (Fig. 3.1b) similar to that seen in many insect groups. These small gastropods and other members within their subclass are hypothesized to have evolved from shelled snails into sea slugs as they developed the ability to feed on defended prey and to sequester the prey's defenses (Faulkner and Ghiselin 1983). As examples, the eolid nudibranchs sequester stinging capsules from their coelenterate prey; the dorid nudibranchs sequester sponge toxins; and the ascoglossans sequester both defensive compounds and functional chloroplasts from seaweeds. Sequestered chloroplasts are retained intracellularly and sustain photosynthesis for up to three months, with the photosynthetic products becoming available to the ascoglossan (Trench 1975). Because energy from chloroplast symbiosis can supply some ascoglossans with all of their respiratory needs, these ascoglossans may do minimal damage to their host plants. For ascoglossans that depend on their host plants to provide food, habitat, and protection from predators in the form of sequesterable metabolites, reduction of damage to the host plant could be adaptive.

Ascoglossans feed by sucking algal sap rather than by chewing. Most species feed on *Caulerpa* spp. or other chemically rich green algae in the families Halimedaceae and Udoteaceae (*Halimeda, Udotea, Penicillus, Chlorodesmis, Avrainvillea,* and others); however, a few species specialize on other foods (Jensen 1980, 1983). Few studies have rigorously assessed feeding by ascoglossans, but natural history observations, laboratory feeding trials, and chemical assays all suggest that most species feed on one or a small number of related algae (Jensen 1980, 1983, 1984; Paul and Van Alstyne 1988b; Hay et al. 1989, 1990b). Although it is widely assumed that ascoglossans are defended from predation by sequestered algal metabolites contained in the cerata and mucus that they release when disturbed, rigorous investigations of this phenomenon are rare.

On Guam, the ascoglossan *Elysia halimedae* occurs exclusively on the green alga *Halimeda* and appears to graze selectively those tissues with the highest concentrations of chemical defenses (Paul and Van Alstyne 1988b). Once consumed, the major metabolite of *Halimeda*, halimedatetraacetate (8) is converted to a related diterpene alcohol (9)

8 R = CHO
9 R = CH$_2$OH

that the ascoglossan sequesters as up to 7% of its dry mass. The compound is also deposited in the ascoglossan's egg masses. In field assays, a concentration of 4% of this compound significantly deterred feeding by both herbivorous and carnivorous fishes (Paul and Van Alstyne 1988b).

In the Caribbean, the ascoglossan *Costasiella ocellifera* is found only on the green alga *Avrainvillea*. This alga produces the brominated diphenylmethane derivative, avrainvilleol (6) (Sun et al. 1983), which deters fish from grazing and is sequestered by the ascoglossan (Hay et al. 1990b). The ascoglossan was rejected unharmed when offered to the predatory wrasse *Thalassoma bifasciatum*, and both the organic crude extract of the ascoglossan and the pure compound avrainvilleol significantly deterred feeding by this same fish when these substances were injected into palatable foods. Somewhat similar interactions may occur with ascoglossans that feed on the chemically defended (Hay et al. 1987a) Caribbean alga *Cymopolia barbata* (Jensen 1984) and on the chemically defended (Paul 1987; Wylie and Paul 1988) Pacific alga *Chlorodesmis fastigiata* (Hay et al. 1989); however, the chemistry of these interactions has not been adequately documented.

Ascoglossans sometimes contain polypropionate compounds that are not related to metabolites in their algal foods (Dawes and Wright 1986; Roussis et al. 1990). Gastropods appear to synthesize these compounds de novo (Ireland and Faulkner 1981; Manker et al. 1988). Although the ecological function of the polypropionates is unknown, limited assays suggest that they do not play a defensive role against fishes (Roussis et al. 1990; J. R. Pawlik, pers. comm., 1991).

Like ascoglossans, sea hares sequester defensive compounds from chemically rich seaweeds; but unlike the ascoglossans, these shell-less gastropods are not restrictively specialized feeders. They sequester metabolites from the blue-green alga *Lyngbya majuscula*, the brown algae *Dictyota*, *Dictyopteris*, and *Glossophora*, and the red algae *Laurencia*, *Plocamium*, and *Asparagopsis* (Faulkner 1984). Sea hares preferentially consume the chemically rich seaweeds listed above but will also eat a wide variety of algae that do not contain unusual metabolites (Carefoot 1987). Their large size and rapid growth rate (Audesirk 1979; Carefoot 1987) may preclude restrictive specialization to a single prey species because an adult would consume many plants each day, rapidly depleting its local food supply and forcing its migration to other host patches. The patchiness of preferred foods and the limited mobility of gastropods makes specialization an improbable strategy. Juvenile *Aplysia californica* were thought to selectively settle on and consume species of *Laurencia* and *Plocamium*, which produce toxic secondary metabolites that are sequestered by the sea hare and then used in its own defense (Kriegstein et al. 1974; Kupfermann and Carew 1974; Stallard and Faulkner 1974; Norris and Fenical 1982; Pennings 1991). More recent work suggests that *Aplysia* larvae may settle when they encounter any of numerous species of red algae even though the juveniles then selectively crawl to and eat only a limited number of the plants (Pawlik 1989; Pennings 1990a, 1990b).

Since *Aplysia* appear to be able to grow and reproduce equally well on either *Plocamium* or *Enteromorpha* (a genus without chemical defenses; Carefoot 1967), the preference for chemically rich seaweeds may reflect the need for predator defense rather than some unique dietary requirement. When small *Aplysia californica* were grown on either the terpene-rich alga *Plocamium* or the terpene-free alga *Ulva*, those grown on *Plocamium* were rich in terpenes and much less susceptible to wrasse predation than those grown on *Ulva*, which were terpene free (Pennings 1990a); however, lobsters and several fish species readily ate *Aplysia* grown on either diet.

When *Aplysia* consume chemically rich seaweeds, the algal compounds are sequestered in the digestive gland (Willcott et al. 1973; Stallard and Faulkner 1974; Kinnel et al. 1979). In *A. californica*, the seaweed-derived polyhalogenated terpenes and sesquiterpenes stored in this gland make up 0.7% of the animal's wet mass and have a turnover time of less than three months (Stallard and Faulkner 1974). These compounds are slowly moved from the digestive gland to the skin (Willcott et al. 1973; Kinnel et al. 1979), where they may function as predator deterrents. When disturbed, *Aplysia* produce a purple ink that is not known to be toxic; if the disturbance continues, they produce a milky fluid containing toxic oils that are reported to cause muscular paralysis and death of some marine organisms (Stallard and Faulkner 1974). Although sea hares are not restrictively specialized to a single host plant, they are clearly more specialized than most marine herbivores in that they selectively eat and sequester defensive metabolites from certain chemically rich seaweeds.

Chemically Mediated Associational Refuges and Defenses

Just as defended seaweeds can serve as safe sites and sources of defensive resources for small herbivores, they may also provide safe sites for palatable seaweeds (Hay 1985, 1986; Littler et al. 1986; Pfister and Hay 1988). There is an implied chemical nature to these interactions, but this has not been rigorously investigated.

In the Caribbean, the chemically defended (Hay et al. 1987b) alga *Stypopodium zonale* produces an associational refuge for other, more palatable, species; several palatable seaweeds are significantly more common near the base of *Stypopodium* than several centimeters away (Littler et al. 1986). When *Stypopodium* plants were experimentally removed, palatable seaweeds nearby experienced greater losses to grazers than did palatable species near the bases of control (i.e., not removed) *Stypopodium* plants. When plastic mimics of *Stypopodium* were placed in the field, palatable plants near these mimics experienced significantly reduced grazing compared with palatable plants placed in the open; but there was even less grazing near real *Stypopodium* plants (Littler et al. 1986). These findings suggest that a portion of the associational refuge is generated by the mere physical presence of a nonfood plant, but the chemical repugnance of the unpalatable plant may provide additional safety to the palatable species. Similar positive associations occur between palatable seaweeds and the unpalatable sea fan *Gorgonia ventalina* and fire coral *Millepora alcicornis* (Littler et al. 1987).

In coastal North Carolina, palatable red and green algae gain significant protection from herbivorous fishes by growing epiphytically on the unpalatable brown alga *Sargassum filipendula* (Hay 1986). When herbivorous fishes were excluded, palatable species growing on *Sargassum* grew much slower than those growing alone. *Sargassum* is therefore a competitor that suppresses the growth of these palatable seaweeds. When herbivores were present, however, palatable species appeared to depend completely on their unpalatable competitors to provide microsites of reduced herbivory that prevented grazers from driving the palatable seaweeds to local extinction. Thus the cost of being associated with an unpalatable competitor was much less than the cost of increased herbivory in the absence of that competitor. Under these conditions one competitor can have a strong positive effect on another, and associational refuges can provide an unappreciated mechanism for maintaining species richness within communities dominated by a few unpalatable species (Hay 1986). Although escapes provided by growing on *Sargassum* were initially interpreted as arising from simple visual crypsis (Hay 1986), more recent and detailed investigations of associational refuges from grazing sea urchins suggest that chemistry may play a significant role (Pfister and Hay 1988).

As with the example of *Sargassum* and grazing fishes described above, the palatable red alga *Gracilaria tikvahiae* experiences significantly less grazing by the sea urchin *Arbacia punctulata* when it occurs in the understory of *Sargassum* than when it occurs alone (Pfister and Hay 1988). Experiments with different types of plastic *Sargassum* mimics indicated that the decreased grazing on *Gracilaria* was not a result of *Sargassum* morphology or of simply including nonfood items that interfered with urchin movement. The experiments suggested, but did not conclusively prove, that some aspect of *Sargassum* chemistry decreased the foraging and feeding activity of urchins in the immediate vicinity of *Sargassum*.

Increased grazing in monocultures of host plants versus polycultures containing some nonhost plants also has been demonstrated for plants and herbivorous insects in terrestrial communities (Tahvanainen and Root 1972; Root 1973, 1974; Atsatt and O'Dowd 1976). In these communities, insect density is higher in monocultures than in polycultures because insects immigrate to monocultures more rapidly, emigrate from them less often, and experience increased reproduction while in monocultures (Bach 1980; Karieva 1982). In some cases, nonhost terrestrial plants give off chemical stimuli that appear to interfere with the ability of herbivorous insects to find and feed on the host (Tahvanainen and Root 1972). This was not the case for the sea urchin–algal interactions studied by Pfister and Hay (1988). In their experiments, urchins migrated to *Gracilaria* alone or *Gracilaria* growing with *Sargassum* at equal rates. Furthermore, urchins tended to spend more time in the polycultures than in the monocultures *Gracilaria* monocultures. Thus, although the patterns in marine and terrestrial systems were very similar, the mechanisms producing the patterns were strikingly different. The increased grazing in *Gracilaria* monocultures resulted from increased rates of movement and feeding by individual herbivores, not from increased herbivore density, as had been reported for terrestrial communities. The data from this study are consistent with the hypothesis that *Sargassum* produces a compound that does not repel urchins but decreases their activity and feeding rates. Rigorous chemical investigations were not conducted, however, and no compound producing such an effect was isolated or identified. The above-mentioned studies as well as investigations of zooplankton feeding among chemically defended phytoplankton (Egloff 1986; Huntley et al. 1986; Fulton and Paerl 1987) all suggest that further studies of the potential chemical mediation of these interactions would be profitable.

The associational escapes discussed above are opportunistic. They do not appear to be coevolved or to require a long history of co-occurrence of the participants (Hay 1986). There are, however, other types of associational interactions that may be coevolved. If these largely unstudied associations are more predictable and the participants more tightly coupled, then these may be active associational *defenses* rather than passive associational *refuges*. As a potential example, on the coast of Oregon the green alga *Enteromorpha vexata* was eaten to local extinction by grazing gastropods unless it was infected by the ascomycete fungus *Turgidosculum*. Infected plants were the only green algae that remained abundant during periods of maximal herbivore effectiveness, and uninfected plants were preferred over infected plants by the snail *Littorina scutulata* (Cubit 1975). Although the chemistry involved in this interaction has not been studied, the relationship appears similar to well-studied cases where grasses and other plants infected with ascomycete fungi were protected from grazers, parasites, and microbial pathogens through alkaloids produced by the fungi (Carroll 1988; Cheplick and Clay 1988; Clay 1988). In addition to the immediate advantages of acquiring chemical defenses from a fungal symbiont, there could be significant evolutionary advantages to this relationship if chemical defenses of the symbiont evolved more rapidly than those based in the host genome (Carroll 1988).

Since almost one-third of all known higher marine fungi are associated with algae (Kohlmeyer and Kohlmeyer 1979), fungal-generated chemical defenses could be important in marine as well as terrestrial systems. In marine systems these associations range from parasitic to symbiotic and from facultative to obligate. Kohlmeyer and Kohlmeyer (1979) listed a large number of marine fungi and their algal hosts. Obligate symbiotic examples included *Blodgettia bornetii* and green algae in the genus *Cladophora*, *Mycosphaerella ascophylli* and the brown algae *Ascophyllum nodosum* and *Pelvetia canaliculata*, *Turgidosculum complicatulum* and green algae in the genus *Prasiola*, and *T. ulvae* and the green alga *Blidingia minima* var. *vexata* (= *Ulva vexata*).

I know of only one study that rigorously addressed how symbiotic microbes chemically mediate host-enemy interactions. Embryos of the shrimp *Palaemon macrodactylus* are consistently covered by a strain of the bacterium *Alteromonas* sp. that produces 2,3-indolinedione (isatin) (10),

10

which defends the embryos from attack by the pathogenic fungus *Lagenidium callinectes* (Gil-Turnes et al. 1989). When the bacterium was removed from the embryos, they suffered greater than 90% mortality when exposed to the fungus. If the bacterium was removed and then replaced, or removed and the embryos treated with the pure compound isatin, embryo survivorship was not significantly different from embryos with unmanipulated bacterial populations. Although relatively little is known about them, such interactions could be common in marine systems. Many benthic invertebrates and seaweeds support populations of symbiotic, commensal, or parasitic organisms that could play important roles through chemical mediation of significant biotic interactions. The potential complexity of such interactions and the difficulty of clearly documenting the role of chemical mediation may make these investigations difficult, but terrestrial research offers some insight into how these problems could be approached and to their potential importance (Price et al. 1986).

Tolerating Herbivory

Rather than escaping or deterring herbivores, some plants may persist in herbivore-affected habitats by tolerating herbivory and evolving ways to minimize the negative effects of being eaten. In some cases, tolerance may be affected by chemical characteristics of the algae. As an example, colonies of the green alga *Sphaerocystis schroeteri* consist of cells embedded in a complex polysaccharide sheath. When the alga was consumed by *Daphnia magna, D. galeata*, and other natural predators, fewer than 10% of the cells were damaged during gut passage (Porter 1976). The nutrient enrichment that occurred during gut passage stimulated the growth rate by as much as 63%, more than compensating for the slight damage due to grazing. Grazers therefore provided this alga with a rich source of localized nutrients. In situ grazing experiments have shown that the density of this alga increases as grazer density increases.

Somewhat similar interactions can occur between grazers and spores of benthic seaweeds. When some adult seaweeds are eaten by herbivores, gut passage can significantly increase the production of motile spores and the growth rate of sporelings relative to uningested controls (Buschmann and Santelices 1987; Santelices and Ugarte 1987). In these situations the seaweeds are dispersed by the herbivores and minimize the damage due to herbivory by producing spores that survive gut passage and apparently absorb nutrients during the process. A chemical basis for resistance to digestion has not been investigated but is an obvious possibility.

Interactions among Seaweed Defensive Metabolites

Many seaweeds produce several related secondary metabolites. Within such seaweeds, these multiple compounds could act additively or synergistically or could reduce the ability of herbivores to develop resistance to seaweed defenses. Studies of seaweed chemical defense have concentrated on determining the effects of individual metabolites; the interactions and potentially complex effects of multiple secondary metabolites have not been experimentally addressed. Within this context, it should be noted that metabolites shown to be ineffective in deterring herbivores when tested alone could still play a defensive role if they change the ability of active compounds to reach their molecular target site or if they limit the ability of herbivores to block the action of effective compounds. Compounds playing such synergistic roles are known to occur in terrestrial plants; Berenbaum (1985) provided several examples and discussed the range of complex interactions known to occur among allelochemicals in terrestrial plants.

Additional Effects of Chemical Defense

Although numerous seaweed secondary metabolites have been demonstrated to significantly deter herbivores in both field and laboratory assays (Hay and Fenical 1988; Hay 1991a), these compounds could serve additional functions as antifouling agents, allelopathic agents, and so forth. These possible alternate functions have not received adequate attention.

Furthermore, since secondary metabolites are often strongly bioactive and may not degrade immediately after the death of the plant, these compounds could significantly affect ecosystem-level processes such as carbon and nutrient cycling. As examples, (1) phenolic acids in the marsh grass *Spartina* take several months to leach from dead plant material, and at natural concentrations they significantly deter feeding by salt-marsh detritivores (Valiela et al. 1979; Valiela and Rietsma 1984; Wilson et al. 1986; Rietsma et al. 1988); and (2) polyphenolics from algae can affect the characteristics of near-shore waters by affecting formation of humic materials (Sieburth and Jensen 1969) and chelation of ions (Ragan and Glombitza 1986). Thus secondary metabolites may not only deter herbivores but also may determine chemical characteristics of the water column and the time at which detritus becomes palatable to decomposers. If microbial activity is also inhibited by secondary metabolites, this may further slow the degradation of plant detritus. Effects of plant metabolites on decomposition have been more extensively studied in terrestrial communities (Horner et al. 1988).

The ecosystem-wide effects of phytoplankton secondary metabolites (e.g., red tides) can be dramatic (Paerl 1988; Steidinger and Vargo 1988). These effects include large-scale die-offs of fishes, scallops, and other nontarget organisms; the loss of millions of dollars in fisheries revenues and thousands of tons of seafood harvest; and occasionally, respiratory impairment of terrestrial animals inhabiting coastal areas.

Outbreaks of the dinoflagellate *Ptychodiscus brevis* along the Gulf Coast of the United States may kill 100 tons of fish per day, and similar effects have been reported in California, Japan, Korea, India, Sri Lanka, and North Carolina (Paerl 1988). In Scandinavia, a recent outbreak of the gelatinous alga *Chrysochromulina polylepsis* cost the Norwegian fishing industry approximately $200 million and included the loss of 500 tons of salmon (Cotter 1988).

On an even broader scale, it has recently been argued that bromoforms, and possibly other haloforms, produced by the breakdown of red and brown marine algae may significantly affect depletion of the earth's ozone layer (Barrie et al. 1988; Weaver 1988). This may help explain the regular springtime occurrences of ozone depletion at ground level in a 10-year data base from Barrow, Alaska, as well as the peaks in aerosol bromine observed throughout the Arctic in March and April (Barrie et al. 1988). Given the strong chemical and biological activities of many seaweed metabolites, it may be common for them to produce significant indirect effects, many of which could function through complex mechanisms. Future investigations in these areas should be productive.

Acknowledgments Charles Birkeland, J. Emmett Duffy, William Fenical, Valerie Paul, Peter Steinberg, and Robbin Trindell provided comments that improved the manuscript. My recent work on seaweed chemical defenses has been funded by U.S. National Science Foundation grants OCE 89-11872 and OCE 89-00131, the National Geographic Society, and the Lizard Island Reef Research Foundation of the Australian Museum.

References

Atsatt, P.R., and O'Dowd, D.J. 1976. Plant defense guilds. Science 193:24–29.
Audesirk, T.E. 1979. A field study of growth and reproduction of *Aplysia californica*. Biol. Bull. 157:407–421.
Bach, C.E. 1980. Effects of plant diversity and time of colonization on an herbivore-plant interaction. Oecologia 44:319–326.
Barbosa, P. 1988. Some thoughts on "the evolution of host range." Ecology 69:912–915.
Barrie, L.A., Bottenheim, J.W., Schnell, R.C., Crutzen, P.J., and Rasmussen, R.A. 1988. Ozone destruction and photochemical reactions at polar sunrise in the lower Arctic atmosphere. Nature 334:138–141.
Berenbaum, M. 1985. Brementown revisited: interactions among allelochemicals in plants. Recent Adv. Phytochem. 19:139–169.

Bernays, E.A. 1989. Host range in phytophagous insects: the potential role of generalist predators. Evol. Ecol. 3:299–311.

Bernays, E., and Cornelius, M.L. 1989. Generalist caterpillar prey are more palatable than specialists for the generalist predator *Iridomyrmex humilus*. Oecologia 79:427–430.

Bernays, E., and Graham, M. 1988. On the evolution of host specificity in phytophagous arthropods. Ecology 69:886–892.

Buschmann, A., and Santelices, B. 1987. Micrograzers and spore release in *Iridaea laminarioides* Bory (Rhodophyta: Gigartinales). J. Exp. Mar. Biol. Ecol. 108:171–179.

Carefoot, T.H. 1967. Growth and nutrition of *Aplysia punctata* feeding on a variety of marine algae. J. Mar. Biol. Assoc. U.K. 47:565–589.

Carefoot, T.H. 1987. *Aplysia*: its biology and ecology. Oceanogr. Mar. Biol. Annu. Rev. 25:167–284.

Carpenter, R.C. 1986. Partitioning herbivory and its effects on coral reef algal communities. Ecol. Monogr. 56:345–363.

Carroll, G. 1988. Fungal endophytes in stems and leaves: from latent pathogen to mutualistic symbiont. Ecology 69:2–9.

Chelplik, G.P., and Clay, K. 1988. Acquired chemical defenses in grasses: the role of fungal endophytes. Oikos 52:309–318.

Clark, K.B., and DeFreeze, D. 1987. Population ecology of Caribbean Ascoglossa (Mollusca: Opisthobranchia): a study of specialized algal herbivores. Am. Malacol. Bull. 5:259–280.

Clay, K. 1988. Fungal endophytes of grasses: a defensive mutualism between plants and grasses. Ecology 69:10–16.

Cotter, R. 1988. Scandinavian killer algae outbreak. Nature 333:488.

Courtney, S. 1988. If it's not coevolution, it must be predation? Ecology 69:910–911.

Cubit, J.D. 1975. Interactions of seasonally changing physical factors and grazing affecting high intertidal communities on a rocky shore. Ph.D. dissertation, University of Oregon, Eugene.

Damman, H. 1987. Leaf quality and enemy avoidance by the larvae of a pyralid moth. Ecology 68:88–97.

Dawes, R.D., and Wright, J.L.C. 1986. The major propionate metabolites from the sacoglossan mollusc *Elysia chlorotica*. Tetrahedron Lett. 27:255–258.

Dayton, P.K., and Tegner, M.J. 1990. Bottoms below troubled waters: benthic impacts of the 1982–1984 El Niño in the temperate zone. *In* Global ecological consequences of the 1982–1983 El Niño—southern oscillation, ed. P. W. Glynn, pp. 433–472. Amsterdam: Elsevier Oceanography Series.

Duffy, J.E. 1989. Ecology and evolution of herbivory by marine amphipods. Ph.D. dissertation, University of North Carolina at Chapel Hill.

Duffy, J.E., and Hay, M.E. 1990. Seaweed adaptations to herbivory. BioScience 40:368–375.

Duffy, J.E., and Hay, M.E. 1991a. Amphipods are not all created equal: a reply to Bell. Ecology 72:354–358.

Duffy, J.E., and Hay, M.E. 1991b. Food and shelter as determinants of food choice by an herbivorous marine amphipod. Ecology 72:1286–1298.

Edgar, G.J. 1983a. The ecology of southeast Tasmanian phytal animal communities. II. Seasonal changes in plant and animal populations. J. Exp. Mar. Biol. Ecol. 70:159–179.

Edgar, G. J. 1983b. The ecology of southeast Tasmanian phytal animal communities.

IV. Factors affecting the distribution of ampithoid amphipods among algae. J. Exp. Mar. Biol. Ecol. 70:205–225.

Egloff, D.A. 1986. Effects of *Olisthodiscus luteus* on the feeding and reproduction of the marine rotifer *Synchaeta cecilia*. J. Plankton Res. 8:263–274.

Ehrlich, P.R., and Murphy, D.D. 1988. Plant chemistry and host range in insect herbivores. Ecology 69:908–909.

Estes, J.A., and Steinberg, P.D. 1988. Predation, herbivory, and kelp evolution. Paleobiology 14:19–36.

Faulkner, J.D. 1984. Marine natural products: metabolites of marine algae and herbivorous marine molluscs. Nat. Prod. Rep. 1:251–280.

Faulkner, J.D., and Ghiselin, M.T. 1983. Chemical defense and evolutionary ecology of dorid nudibranchs and some other opisthobranch gastropods. Mar. Ecol. Prog. Ser. 13:295–301.

Fox, L.R. 1981. Defense and dynamics in plant-herbivore interactions. Am. Zool. 21:853–864.

Fox, L.R. 1988. Diffuse coevolution within complex communities. Ecology 69:906–907.

Fulton, R.S., III, and Paerl, H.W. 1987. Toxic and inhibitory effects of the blue-green alga *Microcystis aeruginosa* on herbivorous zooplankton. J. Plankton Res. 9:837–855.

Futuyma, D.J., and Gould, F. 1979. Associations of plants and insects in a deciduous forest. Ecol. Monogr. 49:33–50.

Gilbert, L.E., and Singer, M.C. 1975. Butterfly ecology. Annu. Rev. Ecol. Syst. 6:365–397.

Gil-Turnes, M.S., Hay, M.E., and Fenical, W. 1989. Symbiotic marine bacteria chemically defend crustacean embryos from a pathogenic fungus. Science 246:116–118.

Hawkins, S.J., and Hartnoll, R.G. 1983. Grazing of intertidal algae by marine invertebrates. Oceanogr. Mar. Biol. Annu. Rev. 21:195–282.

Hay, M.E. 1984. Predictable spatial escapes from herbivory: how do these affect the evolution of herbivore resistance in tropical marine communities? Oecologia 64:396–407.

Hay, M.E. 1985. Spatial patterns of herbivore impact and their importance in maintaining algal species richness. *In* Proc. Fifth Int. Coral Reef Symp., vol. 4, pp. 29–34.

Hay, M.E. 1986. Associational plant defenses and the maintenance of species diversity: turning competitors into accomplices. Am. Nat. 128:617–641.

Hay, M.E. 1991a. Fish-seaweed interactions on coral reefs: effects of herbivorous fishes and adaptations of their prey. *In* The ecology of fishes on coral reefs, ed. P.F. Sale, pp. 96–119. San Diego: Academic Press.

Hay, M.E. 1991b. Marine-terrestrial contrasts in the ecology of plant chemical defenses against herbivores. TREE 6:362–364.

Hay, M.E., Duffy, J.E., and Fenical, W. 1988d. Seaweed chemical defenses: among-compound and among-herbivore variance. *In* Proc. Sixth Int. Coral Reef Symp., vol. 3, pp. 43–48.

Hay, M.E., Duffy, J.E., and Fenical, W. 1990a. Host-plant specialization decreases predation on a marine amphipod: an herbivore in plant's clothing. Ecology 71:733–743.

Hay, M.E., Duffy, J.E., Fenical, W., and Gustafson, K. 1988a. Chemical defense in the seaweed *Dictyopteris delicatula*: differential effects against reef fishes and amphipods. Mar. Ecol. Prog. Ser. 48:185–192.

Hay, M.E., Duffy, J.E., Paul, V.J., Renaud, P.E., and Fenical, W. 1990b. Specialist herbivores reduce their susceptibility to predation by feeding on the chemically-defended seaweed *Avrainvillea longicaulis*. Limnol. Oceanogr. 35:1734–1743.

Hay, M.E., Duffy, J.E., Pfister, C.A., and Fenical, W. 1987a. Chemical defenses against different marine herbivores: are amphipods insect equivalents? Ecology 68:1567–1580.

Hay, M.E., and Fenical, W. 1988. Marine plant-herbivore interactions: the ecology of chemical defense. Annu. Rev. Ecol. Syst. 19:111–145.

Hay, M.E., Fenical, W., and Gustafson, K. 1987b. Chemical defense against diverse coral-reef herbivores. Ecology 68:1581–1591.

Hay, M.E., and Fuller, P.J. 1981. Seed escape from heteromyid rodents: the importance of microhabitat and seed preference. Ecology 62:1395–1399.

Hay, M.E., Paul, V.J., Lewis, S.M., Gustafson, K., Tucker, J., and Trindell, R.N. 1988b. Can tropical seaweeds reduce herbivory by growing at night? Diel patterns of growth, nitrogen content, herbivory, and chemical versus morphological defenses. Oecologia 75:233–245.

Hay, M.E., Pawlik, J.R., Duffy, J.E., and Fenical, W. 1989. Seaweed-herbivore-predator interactions: host-plant specialization reduces predation on small herbivores. Oecologia 81:418–427.

Hay, M.E., Renaud, P.E., and Fenical, W. 1988c. Large mobile versus small sedentary herbivores and their resistance to seaweed chemical defenses. Oecologia 75:246–252.

Hay, M.E., and P.D. Steinberg. 1992. The chemical ecology of plant-herbivore interactions in marine versus terrestrial communities. *In* Herbivores: their interaction with secondary plant metabolites. Vol. 2: Ecological and evolutionary processes, ed. G.A. Rosenthal and M. Berenbaum. San Diego: Academic Press, in press.

Holbrook, S.J., and Schmitt, R.J. 1988. The combined effects of predation risk and food reward on patch selection. Ecology 69:125–134.

Horner, J.D., Gosz, J.R., and Cates, R.G. 1988. The role of carbon based secondary metabolites in decomposition in terrestrial ecosystems. Am. Nat. 132:869–883.

Huntley, M., Sykes, P., Rohan, S., and Martin, V. 1986. Chemically-mediated rejection of dinoflagellate prey by the copepods *Calanus pacificus* and *Paracalanus parvus*: mechanism, occurrence and significance. Mar. Ecol. Prog. Ser. 28:105–120.

Ireland, C., and Faulkner, D.J. 1981. The metabolites of the marine molluscs *Tridachiella diomedea* and *Tridachia crispata*. Tetrahedron 37:233–240.

Janzen, D.H. 1988. On the broadening of insect-plant research. Ecology 69:905.

Jensen, K.R. 1980. A review of sacoglossan diets, with comparative notes on radular and buccal anatomy. Malacol. Rev. 13:55–77.

Jensen, K.R. 1983. Factors affecting feeding selectivity in herbivorous ascoglossa (Mollusca: Opisthobranchia). J. Exp. Mar. Biol. Ecol. 66:135–148.

Jensen, K.R. 1984. Defensive behavior and toxicity of ascoglossan opisthobranch *Mourgona germaineae* Marcus. J. Chem. Ecol. 10:475–486.

Jermy, T. 1988. Can predation lead to narrow food specialization in phytophagous insects? Ecology 69:902–904.

Jones, L.G. 1965. Canopy grazing at southern Point Loma. *In* Kelp Habitat Improvement Project annual report, Feb. 1964–31 Mar. 1965, vol. 1, pp. 62–63. W.M. Keck Laboratory of Environmental Health and Engineering. Pasadena: California Institute of Technology.

Karieva, P. 1982. Experimental and mathematical analyses of herbivore movement: quantifying the influence of plant spacing and quality on foraging discrimination. Ecol. Monogr. 52:261–282.

Kawamura, K. 1973. Fisheries biological studies on a sea urchin, *Strongylo-centrotus intermedius* (A. Agassiz). Sci. Rep. Hokkaido Fish. Exp. Stn. 16:1–54.

Kinnel, R.B., Dieter, R.K., Meinwald, J., Van Engen, D., Clardy, J., Eisner, T., Stallard, M.O., and Fenical, W. 1979. Brasilenyne and *cis*-dihydrorhodophytin: antifeedant medium-ring haloesters from a sea hare (*Aplysia brasiliana*). Proc. Natl. Acad. Sci. USA 76:3576–3579.

Kohlmeyer, J., and Kohlmeyer, E. 1979. Marine mycology: the higher fungi. New York: Academic Press.

Kriegstein, A.R., Castellucci, V., and Kandell, E.R. 1974. Metamorphosis of *Aplysia californica* in laboratory culture. Proc. Natl. Acad. Sci. USA 71:3654–3658.

Kupfermann, I., and Carew, T.J. 1974. Behavior patterns of *Aplysia californica* in its natural environment. Behav. Biol. 12:317–337.

Lawton, J.H. 1978. Host-plant influences on insect diversity: the effects of space and time. *In* Diversity of insect faunas. Symposia of the Royal Entomological Society of London, vol. 9, ed. L.A. Mounds and N. Waloff, pp. 105–125. Oxford: Blackwell.

Lewis, S.M. 1986. The role of herbivorous fishes in the organization of a Caribbean reef community. Ecol. Monogr. 56:183–200.

Lewis, S.M., and Kensley, B. 1982. Notes on the ecology and behavior of *Pseudamphithoides incurvaria* (Just) (Crustacea, Amphipoda, Ampithoidae). J. Nat. Hist. 16:267–274.

Lewis, S.M., Norris, J.N., and Searles, R.B. 1987. The regulation of morphological plasticity in tropical reef algae by herbivory. Ecology 68:636–641.

Littler, M.M., Littler, D.S., and Taylor, P.R. 1987. Animal-plant defense associations: effects on the distribution and abundance of tropical reef macrophytes. J. Exp. Mar. Biol. Ecol. 105:107–121.

Littler, M.M., Taylor, P.R., and Littler, D.S. 1983. Algal resistance to herbivory on a Caribbean barrier reef. Coral Reefs 2:111–118.

Littler, M.M., Taylor, P.R., and Littler, D.S. 1986. Plant defense associations in the marine environment. Coral Reefs 5:63–71.

Lowe, E.F. 1975. Absorption efficiencies, feeding rates, and food preferences of *Lytechinus variegatus* (Echinodermata: Echinoidea) for selected marine plants. M.Sc. thesis, University of South Florida, Tampa.

Lubchenco, J., and Gaines, S.D. 1981. A unified approach to marine plant-herbivore interactions. I. Populations and communities. Annu. Rev. Ecol. Syst. 12:405–437.

Manker, D.C., Garson, M.J., and Faulkner, D.J. 1988. *De novo* biosynthesis of polypropionate metabolites in the marine pulmonate *Siphonaria denticulata*. J. Chem. Soc. Chem. Comm. 1988:1061–1062.

Morrison, D. 1988. Comparing fish and urchin grazing in shallow and deeper coral reef algal communities. Ecology 69:1367–1382.

Nelson, W.G. 1979a. An analysis of structural pattern in an eelgrass (*Zostera marina* L.) amphipod community. J. Exp. Mar. Biol. Ecol. 39:231–264.

Nelson, W.G. 1979b. Experimental studies of selective predation on amphipods: consequences for amphipod distribution and abundance. J. Exp. Mar. Biol. Ecol. 38:225–245.

Norris, J.N., and Fenical, W. 1982. Chemical defense in tropical marine algae. *In* Atlantic barrier reef ecosystem Carrie Bow Cay, Belize, vol. 1, Structure and communities, ed. K. Rutzler and I. G. Macintyre, pp. 417–431. Smithson. Contrib. Mar. Sci. 12.

Ogden, J.C., Abbott, D.P., and Abbott, I.A., eds. 1973. Studies on the activity and food of the echinoid *Diadema antillarum* Philippi on a West Indian patch reef. Spec.

Publ. 2, Fairleigh Dickinson University West Indies Laboratory, St. Croix, Virgin Islands.

Paerl, H.W. 1988. Nuisance phytoplankton blooms in coastal, estuarine, and inland waters. Limnol. Oceanogr. 33:823–847.

Paul, V.J. 1987. Feeding deterrent effects of algal natural products. Bull. Mar. Sci. 41:514–522.

Paul, V.J., and Fenical, W. 1987. Natural products chemistry and chemical defense in tropical marine algae of the phylum Chlorophyta. In Bioorganic marine chemistry, vol. 1, ed. P.J. Scheuer, pp. 1–29. Berlin: Springer-Verlag.

Paul, V.J., and Hay, M.E. 1986. Seaweed susceptibility to herbivory: chemical and morphological correlates. Mar. Ecol. Prog. Ser. 33:255–264.

Paul, V.J., Hay, M.E., Duffy, J.E., Fenical, W., and Gustafson, K. 1987. Chemical defense in the seaweed Ochtodes secundiramea (Montagne) Howe (Rhodophyta): effect of its monoterpenoid components upon diverse coral reef herbivores. J. Exp. Mar. Biol. Ecol. 114:249–260.

Paul, V.J., and Van Alstyne, K.L. 1988a. Chemical defense and chemical variation in some tropical Pacific species of Halimeda (Halimedaceae; Chlorophyta). Coral Reefs 6:263–270.

Paul, V.J., and Van Alstyne, K.L. 1988b. Use of ingested algal diterpenoids by Elysia halimedae Macnae (Opisthobranchia: ascoglossa) as antipredator defenses. J. Exp. Mar. Biol. Ecol. 119:15–29.

Pawlik, J.R. 1989. Larvae of the sea hare Aplysia californica settle and metamorphose on an assortment of macroalgal species. Mar. Ecol. Prog. Ser. 51:195–199.

Pennings, S.C. 1990a. Multiple factors promoting narrow host range in the sea hare Aplysia californica. Oecologia 82:192–200.

Pennings, S.C. 1990b. Size-related shifts in herbivory: specialization in the sea hare Aplysia californica Cooper. J. Exp. Mar. Biol. Ecol. 142:43–61.

Pennings, S.C. 1991. Temporal and spatial variation in the recruitment of the sea hare Aplysia californica, at Santa Catalina Island. In Recent Advances in California Islands Research. Proc. Third Calif. Isl. Symp.

Pfister, C.A., and Hay, M.E. 1988. Associational plant refuges: convergent patterns in marine and terrestrial communities result from differing mechanisms. Oecologia 77:118–129.

Phillips, D.W., and Castori, P. 1982. Defensive responses to predatory seastars by two specialist limpets, Notoacmea insessa (Hinds) and Collisella instabilis (Gould), associated with marine algae. J. Exp. Mar. Biol. Ecol. 59:23–30.

Porter, K.G. 1976. Enhancement of algal growth and productivity by grazing zooplankton. Science 192:1332–1334.

Power, M.E. 1984. Depth distribution of armored catfish: predator-induced resource avoidance? Ecology 65:523–528.

Price, P.W. 1983. Hypotheses on organization and evolution in herbivorous insect communities. In Variable plants and herbivores in natural and managed systems, ed. R.F. Denno and M.S. McClure, pp. 559–598. New York: Academic Press.

Price, P.W., Bouton, C.E., Gross, P., McPherson, B.A., Thompson, J.N., and Weiss, A.E. 1980. Interactions among three trophic levels: influence of plants on interactions between insect herbivores and natural enemies. Annu. Rev. Ecol. Syst. 11:41–65.

Price, P.W., Westoby, M., Rice, B., Atsatt, P.R., Fritz, R.S., Thompson, J.N., and Mobley, K. 1986. Parasite mediation in ecological interactions. Annu. Rev. Ecol. Syst. 17:487–505.

Ragan, M.A., and Glombitza, K.W. 1986. Phlorotannins, brown algal polyphenols. In

Progress in phycological research, vol. 4, ed. F.E. Round and D.J. Chapman, pp. 129–241. Bristol: Biopress.

Randall, J.E. 1967. Food habits of reef fishes of the West Indies. Stud. Trop. Oceanogr. 5:655–897.

Rausher, M.D. 1988. Is coevolution dead? Ecology 69:898–901.

Rietsma, C.S., Valiela, I., and Buchsbaum, R. 1988. Detrital chemistry, growth, and food choice in the salt-marsh snail (*Melampus bidentatus*). Ecology 69:261–266.

Root, R.B. 1973. Organization of a plant-arthropod association in simple and diverse habitats: the fauna of collards (*Brassica oleracea*). Ecol. Monogr. 43:95–124.

Root, R.B. 1974. Some consequences of ecosystem texture. *In* Ecosystem analysis and prediction, ed. S.A. Levin, pp. 83–97. Phila. Soc. Ind. Appl. Math.

Roussis, V., Pawlik, J.R., Fenical, W., and Hay, M.E. 1990. Secondary metabolites of the chemically-rich ascoglossan *Cyerce nigricans*. Experientia 46:327–329.

Santelices, B., and Ugarte, R. 1987. Algal life-history strategies and resistance to digestion. Mar. Ecol. Prog. Ser. 35:267–275.

Schiel, D.R., and Foster, M.S. 1986. The structure of subtidal algal stands in temperate waters. Oceanogr. Mar. Biol. Annu. Rev. 24:265–307.

Schultz, J.C. 1988. Many factors influence the evolution of herbivore diets, but plant chemistry is central. Ecology 69:896–897.

Sieburth, J.M., and Jensen, A. 1969. Studies on algal substances in the sea. II. The formation of gelbstoff (humic material) by exudates of phaeophyta. J. Exp. Mar. Biol. Ecol. 3:275–289.

Sih, A. 1980. Optimal behavior: can foragers balance two conflicting demands? Science 210:1041–1043.

Stallard, M.O., and Faulkner, D.J. 1974. Chemical constituents of the digestive gland of the sea hare *Aplysia californica*. I. Importance of diet. Comp. Biochem. Physiol. B 49:25–35.

Steidinger, K.A., and Vargo, G.A. 1988. Marine dinoflagellate blooms: dynamics and impacts. *In* Algae and human affairs, ed. C.A. Lembi and J.R. Waaland, pp. 373–401. Cambridge: Cambridge University Press.

Steneck, R.S. 1982. A limpet-corallina alga association: adaptations and defenses between a selective herbivore and its prey. Ecology 63:507–522.

Steneck, R.S. 1983. Escalating herbivory and resulting adaptive trends in calcareous algal crusts. Paleobiology 9:44–61.

Stoner, A.W. 1979. Species-specific predation on amphipod Crustacea by the pinfish *Lagodon rhomboides*: mediation by macrophyte standing crop. Mar. Biol. 55:201–207.

Stoner, A.W. 1980. Abundance, reproductive seasonality and habitat preferences of amphipod crustaceans in seagrass meadows of Apalachee Bay, Florida. Contrib. Mar. Sci. 23:63–77.

Strong, D.R., Lawton, J.H., and Southwood, T.R.E. 1984. Insects on plants: community patterns and mechanisms. Oxford: Blackwell Scientific.

Sun, H.H., Paul, V.J., and Fenical, W. 1983. Avrainvilleol. A brominated diphenyl-methane derivative with feeding deterrent properties from the tropical green alga *Avrainvillea longicaulis*. Phytochemistry 22:743–745.

Tahvanainen, J.O., and Root, R.B. 1972. The influence of vegetational diversity on the population ecology of a specialized herbivore, *Phyllotreta cruciferaea* (Coleoptera: Chrysomelidae). Oecologia 10:321–346.

Tegner, M.J., and Dayton, P.K. 1987. El Niño effects on southern California kelp forest communities. Adv. Ecol. Res. 17:243–279.

Thompson, J.N. 1988. Coevolution and alternative hypotheses on insect/plant interactions. Ecology 69:893–895.

Trench, R.K. 1975. Of leaves that crawl: functional chloroplasts in animal cells. *In* Symbiosis, ed. D.H. Jennings and D.L. Lee, pp. 229–265. Cambridge: Cambridge University Press.

Valiela, I., Koumjian, L., Swain, T., Teal, J.M., and Hobbie, J.E. 1979. Cinnamic acid inhibition of detritus feeding. Nature 280:55–57.

Valiela, I., and Rietsma, C.S. 1984. Nitrogen, phenolic acids, and other feeding cues for salt marsh detritivores. Oecologia 63:350–356.

Van Dolah, R.F. 1978. Factors regulating the distribution and population dynamics of the amphipod *Gammarus palustria* in an intertidal salt marsh community. Ecol. Monogr. 48:191–217.

Vince, S., Valiela, I., Backus, N., and Teal, J.M. 1976. Predation by the salt marsh killifish *Fundulus heteroclitus* (L.) in relation to prey size and habitat structure: consequences for prey distribution and abundance. J. Exp. Mar. Biol. Ecol. 23:255–266.

Weaver, R. 1988. Ozone destruction by algae in the Arctic atmosphere. Nature 335:501.

Wells, R.J., and Barrow, K.D. 1979. Acrylic diterpenes containing 3 enol acetate groups from the green alga *Chlorodesmis fastigiata*. Experientia 35:1544–1545.

Werner, E.E., Gilliam, J.F., Hall, D.J., and Mittelbach, G.G. 1983. An experimental test of the effects of predation risk on habitat use in fish. Ecology 64:1540–1548.

Willcott, M.R., Davis, R.E., Faulkner, D.J., and Stallard, M.O. 1973. The configuration and confirmation of 7-chloro-1,6-dibromo-3,7-dimethyl-3,4-epoxy-1-octene. Tetrahedron Lett. 40:3967–3970.

Wilson, J.O., Buchsbaum, R., Valiela, I., and Swan, T. 1986. Decomposition in salt marsh ecosystems: phenolic dynamics during decay of litter of *Spartina alterniflora*. Mar. Ecol. Prog. Ser. 29:177–187.

Wylie, C.R., and Paul, V.J. 1988. Feeding preference of the surgeonfish *Zebrasoma flavescens* in relation to chemical defenses of tropical algae. Mar. Ecol. Prog. Ser. 45:23–32.

Young, D.K., Buzas, M.A., and Young, M.W. 1976. Species densities of macrobenthos associated with seagrass: a field experimental study of predation. J. Mar. Res. 34:577–592.

Young, D.K., and Young, M.W. 1978. Regulation of species densities of seagrass-associated macrobenthos: evidence from field experiments in the Indian River estuary, Florida. J. Mar. Res. 36:569–593.

Chapter 4
Chemical Defenses
of Marine Molluscs

D. JOHN FAULKNER

Most marine molluscs are protected from predation by a hard shell and do not possess any special chemicals that could function in defense. Those molluscs without a shell invariably compensate for the lack of physical protection by employing chemical secretions or nematocyst-based defenses to reduce the incidence of predation. There are some obvious exceptions: most cephalopods lack an external shell, but their great speed and mobility, coupled with the release of "ink," more than compensates for their physical vulnerability. Also, some marine molluscs have both a shell and the ability to produce a chemical exudation, each of which might contribute to the protection of the animals. However, for most shell-less molluscs it has been proposed, though seldom demonstrated, that the animal's survival depends on its use of dietary chemicals. The dietary chemicals are primarily incorporated into defensive secretions that can deter potential predators. Diet-derived chemicals may also provide pigments for camouflage, protect eggs from predation, and, before consumption, act as cues for settling and metamorphosis.

Our current understanding of the importance of a chemical defense mechanism for certain molluscs has resulted from a convergence of studies in two disparate fields. The descriptive studies of biologists and ecologists have often led to the conclusion that the lack of predation on certain marine molluscs must be due to their chemical secretions. Marine natural product chemists have assumed that all sessile or slow-moving marine organisms that did not possess physical

119

defenses must be protected by secondary metabolites. The marriage of these two approaches may seem inevitable, but it has not been easy to accommodate the vastly different philosophies that underlie chemistry and ecology. Philosophical differences between chemists and ecologists have resulted in a vigorous debate about how best to demonstrate the existence of a chemical defense mechanism. This debate has in turn led to the realization that interdisciplinary studies are essential if both disciplines are to be accommodated.

This review concentrates on studies that have considered the defensive role of chemicals isolated from marine molluscs. This narrow scope excludes discussions of mollusc venoms and the many shellfish poisons that are the result of concentration of toxins from microorganisms, particularly dinoflagellates, since these toxins appear to affect human rather than marine predators. To avoid duplication of the review by Joseph Pawlik in this volume, this chapter does not discuss chemical cues for settling and metamorphosis. My primary aim is to evaluate the hypothesis that physically defenseless molluscs require a chemical defense mechanism to survive.

Sea Hares

The biology and ecology of sea hares of the genus *Aplysia* have recently been reviewed in great detail by Carefoot (1987). My review focuses on the ways chemistry influences the biology and ecology of all genera of sea hares. Several lines of evidence indicate that sea hares are protected from predation by the successful deployment of chemicals obtained from their algal diets. Although they are among the largest of the marine molluscs, sea hares are poorly equipped with physical defenses. They have a vestigial shell that is enclosed within the mantle and therefore has little or no defensive value. They are slow moving even when "swimming," and they rarely leave the algal beds on which they feed. Yet no animals are known to prey regularly on them (for a possible exception see Paine 1963), and most of the reported predation seem to be opportunistic, even accidental. It is no wonder that researchers have looked to chemical factors such as pigmentation and the secretions from the skin, ink gland, and opaline gland to explain the lack of predation. It is difficult to assess the relative value of each of these factors, but the concept that the principal defensive chemicals are obtained from dietary sources is attractive because of the very low cost to the sea hare.

It may be a little unconventional to consider camouflage a chemical property, but there is evidence that the background pigmentation of some sea hares is due to deposition of algal pigments. Carefoot (1987) recorded many examples of diet-derived pigmentation, to which I can add two striking examples. Two almost colorless specimens of *Aplysia californica* were found in a salt pond that contained no red algae, while specimens found feeding on *Laurencia obtusa* in the nearby ocean were the normal reddish brown. Even more striking was the observation that very small (<1 cm) specimens of *Aplysia parvula* that were feeding on the red alga *Asparagopsis taxiformis* were the same bright pink as the alga. Older specimens found on nearby *Laurencia johnstonii* were greenish brown, making them difficult to detect among the tufts of algae, but the intermediate-sized animals were easily observed because they showed a gradation of colors from pink to brown. In all cases a pattern of light gray spots was visible against the underlying algal-derived pigmentation. For smaller sea hares, the use of algal pigments to provide camouflage can be a very efficient method to avoid detection—if not predation—as long as the predator depends on visual cues.

The ink gland secretion is the most obvious of the sea hare's defensive secretions, although not all sea hares are capable of producing it. Ink production depends on the presence of red algae in the sea hare's diet. The purple color of the ink is due to a violet pigment called aplysioviolin (1), the monomethyl ester of phycoerythrobilin (2),

1 R = Me   **2** R = H

a pigment characteristic of red algae (Rüdiger 1967a, 1967b; Chapman and Fox 1969). Thus *Aplysia californica*, which preferentially consumes red algae, produces ink while the co-occurring *A. vaccaria*, which feeds on brown algae, does not. The defensive value of the ink is difficult to assess. The sea hare is too slow moving to escape from a predator under the cover of the ink, which is often swept away by wave action in a very short time. It is possible that the ink contains other chemicals that might either narcotize or repel predators (see DiMatteo 1981, 1982), or it may simply act as a warning signal. To define the possible role of ink in the defensive arsenal of the sea hare, it will be necessary to determine the identity of all other chemical constituents of the ink glands.

Very little is known about the nature and function of the opaline gland secretion. The opaline glands are located in the mantle floor and, when stimulated, discharge a milky white secretion into the pallial cavity. The chemical nature of the opaline secretion is unknown, although in some species it is toxic, bitter tasting to humans (!), and strong smelling (Flury 1915). It has been suggested that the "toxins" are of dietary origin, although the evidence supporting this suggestion is weak (Ando 1952, quoted in Carefoot 1987). A report that the opaline secretion may have assisted a juvenile *A. californica* to escape from the tentacles of the sea anemone *Anthopleura xanthogrammica* suggests, albeit inconclusively, a defensive role for the opaline secretion (Carew and Kupfermann 1974).

The role of the digestive (midgut) gland in protecting sea hares is equally perplexing. Even if one accepts that the digestive gland is a repository for toxic and repellant chemicals, its location within the body of the sea hare is not consistent with a defensive role. It is therefore proposed that the chemicals stored in the digestive gland are transported to the skin, from which they are secreted either continuously or in response to attack. There is ample circumstantial evidence to support this hypothesis, but a detailed mechanism for the translocation and deployment of digestive gland constituents has yet to be described.

The digestive gland is the sea hare's largest organ. The chemicals in the digestive gland are directly related to constituents of the diet (for a review see Faulkner 1984a). For example, the chemical constituents of digestive glands of "wild" specimens of *Aplysia californica* consist mainly of halogenated monoterpenes such as (3) from *Plocamium cartilagineum* and halogenated sesquiterpenes such as pacifenol (4)

3   **4**

from *Laurencia pacifica* (Stallard and Faulkner 1974a). When *A. califor-*

nica is fed a single species of *Laurencia* or *Plocamium*, the digestive gland soon contains only those halogenated compounds characteristic of the dietary alga (Stallard and Faulkner 1974a; Pennings 1990). Not all sea hares concentrate halogenated metabolites from red algae. *A. vaccaria* coexists with *A. californica* but feeds on brown algae of the family Dictyotaceae and therefore concentrates the nonhalogenated algal diterpenes, such as crenulide (5), that are typical of that family (Midland et al. 1983). The Hawaiian sea hare *Stylocheilus longicauda* was shown to contain aplysiatoxin (6) (Kato and Scheuer 1974),

5   6

a toxic and irritant metabolite of the cyanophyte *Lyngbya majuscula* (Moore et al. 1984). The most unusual metabolites isolated from sea hares are the dolastatins, very cytotoxic small peptides found in minute quantities in *Dolabella auricularia* (Pettit et al. 1987a, 1987b); the primary source of these compounds is unknown. The extent to which sea hares obtain their digestive gland constituents from dietary algae is illustrated in Table 4.1. The most remarkable aspect of this process of concentrating algal metabolites is the manner in which dietary constituents are sorted into the (presumably distasteful) compounds that are stored in the digestive gland and the metabolites that are catabolized in the normal manner.

It might be a mistake to think that all algal metabolites distasteful to the sea hare are automatically stored in the digestive gland. There is some evidence that the sea hare stores selected algal metabolites in the digestive gland and that these metabolites are the more effective chemical repellents. For example, specimens of *A. parvula* that fed on

Table 4.1

Metabolites of sea hares by class and their algal sources

Sea hare	Chemical class	Algal source[a]	Reference
Aplysia angasi	Brominated sesquiterpene	*Laurencia* sp. (R)	Pettit et al. 1977
	Brominated diterpene	cf. *Laurencia* spp. (R)	Pettit et al. 1978
A. brasiliana	Halogenated C_{15} enynes[b]	cf. *Laurencia* spp. (R)	Kinnel et al. 1977, 1979; Dieter et al. 1979
	Sesquiterpene	*Laurencia obtusa* (R)	Stallard et al. 1978
A. californica	Halogenated monoterpenes	*Plocamium* spp. (R)	Faulkner et al. 1973; Faulkner and Stallard 1973
	Halogenated sesquiterpenes	*Laurencia pacifica* (R)	Ireland et al. 1976; Stallard and Faulkner 1974a
A. dactylomela	Sesquiterpenes	cf. *Laurencia* spp. (R)	Schmitz and McDonald 1974; Schmitz et al. 1978a, 1978b
	Halogenated C_{15} enynes	cf. *Laurencia* spp. (R)	McDonald et al. 1975; Vanderah and Schmitz 1976; Gopichand et al. 1981; Sakai et al. 1986
	Halogenated sesquiterpenes	cf. *Laurencia* spp. (R)	Hollenbeak et al. 1979; Schmitz et al. 1980; González et al. 1983a; Ichiba and Higa 1986; Baker et al. 1988
	Brominated diterpenes	cf. *Laurencia obtusa* (R)	Schmitz et al. 1979, 1982
	Diterpene	cf. *Dictyota* spp. (B)	González et al. 1983b, 1987
	Diterpene quinone	*Stypopodium zonale* (B)	Gerwick and Whatley 1989
A. depilans	Diterpenes	cf. *Dictyota* spp. (B)	Minale and Riccio 1976; Danise et al. 1977
		Pachydictyon coriaceum	
A. kurodai	Halogenated diterpenes	*Laurencia* spp. (R)	Matsuda et al. 1967; Yamamura and Terada 1977
	Diterpene	Brown alga (?)	Miyamoto et al. 1986; Ojika et al. 1990b

Species	Compound	Source (algal)	References
	Elcosanoid dimer	Red alga (?)	Ojika et al. 1990a
	Brominated meroterpenoid	?	Kigoshi et al. 1990
A. limicina	Halogenated monoterpenes	*Plocamium coccineum* (R)	Imperato et al. 1977
A. oculifera	Halogenated C_{15} enynes	*Laurencia* spp. (R)	Schulte et al. 1981; de Silva et al. 1983
A. punctata	Halogenated monoterpenes	*Plocamium coccineum* (R)	Quiñoa et al. 1989
	Elatol (7)	*Laurencia* sp. (R)	Faulkner, pers. observ.
A. vaccaria	Diterpenes	Brown alga	Midland et al. 1983
Bursatella leachii	Aromatic compound (9)	?	Gopichand and Schmitz 1980; Cimino et al. 1987
Dolabella auricularia	Diterpene	Brown alga	Pettit et al. 1976
	Peptides[c]	?	Pettit et al. 1982, 1987a, 1987b, 1989a, 1989b, 1989c, 1990
D. californica	Diterpenes	Brown alga	Ireland et al. 1976; Ireland and Faulkner 1977
Stylocheilus longicauda	Aplysiatoxins (cf. 6)	*Lyngbya majuscala* (C)	Kato and Scheuer 1974; Rose et al. 1978

[a]R = red algae, B = brown algae, C = cyanophyte. "cf." denotes that the compounds in the sea hare are closely related to those found in the algal species listed.
[b]Brasilenyne (14) and *cis* = dihydrorhodophytin (15) are examples of halogenated C_{15} enynes.
[c]The peptides are very minor constituents.

9

a single species of *Laurencia* contained only elatol (7) although the alga contained both elatol and *iso*-obtusol (8)

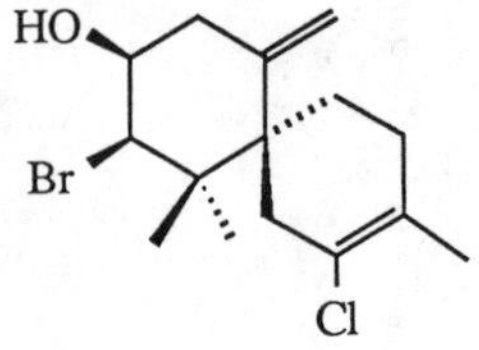

7

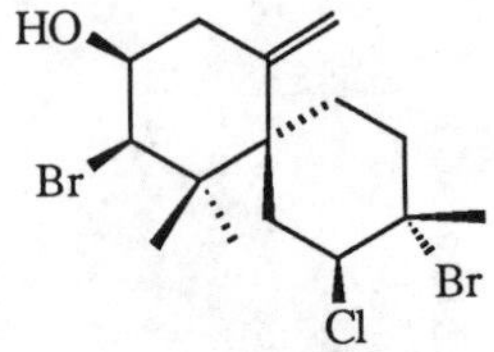

8

in approximately equal proportions (pers. observ.). It appears that sea hares either selectively store elatol or convert *iso*-obtusol into elatol inside the digestive gland. The exact mechanism by which elatol is selectively concentrated is probably less important than the observation that elatol is one of the most cytotoxic halogenated marine natural products known and is a powerful fish feeding deterrent (Hay et al. 1987). While *iso*-obtusol is less active than elatol in many pharmacological assays (antimicrobial, cell division in sea urchin egg), *iso*-obtusol has yet to be compared directly with elatol in palatability assays.

Although the digestive gland serves mainly to store selected algal constituents, it may also be the site of chemical interconversion of these metabolites (Stallard and Faulkner 1974b). Radiolabeling experiments have demonstrated that laurinterol (10), a major constituent of several species of *Laurencia*, was converted into the corresponding cyclic ether, aplysin (11),

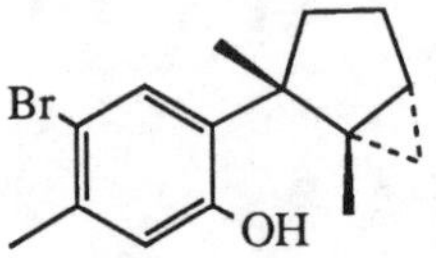

10

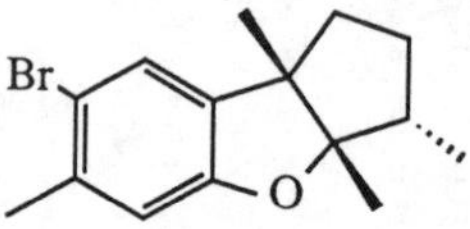

11

in the digestive gland of *A. californica*. This simple conversion, like the proposed conversion of *iso*-obtusol into elatol, could be either a chemical or an enzymatic reaction. The circumstantial evidence slightly favors a chemical reaction because the pH of the digestive gland is acidic and both reactions may be acid catalyzed. Furthermore, there is no firm evidence that the digestive gland is the site of digestive enzymatic activity.

Before the natural product studies were conducted (Table 4.1), several researchers reported that the digestive glands of sea hares contained both lipid-soluble and water-soluble toxins (Winkler 1961; Watson 1973). While the lipid-soluble constituents have been adequately addressed by the natural product studies, the water-soluble toxins have not. It has been suggested, largely on the basis of pharmacological assays, that the water-soluble toxins are derivatives of choline, possibly urocanylcholine (12),

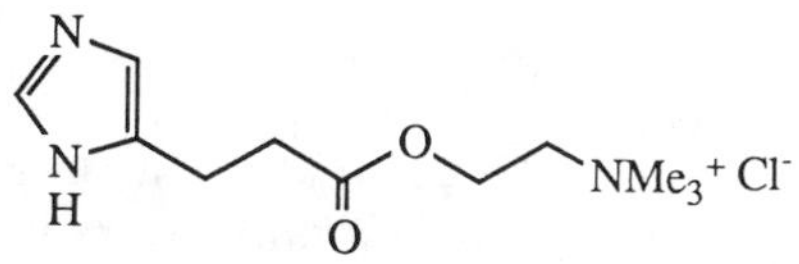

12

but this toxin has not been chemically characterized (Blankenship et al. 1975).

With hindsight it seems obvious that the "bitter" taste of many sea hares is due to their release of chemical feeding deterrents onto the skin. Incorporating these chemicals into a mucous secretion would allow them to be held in a very effective surface coating. Unfortunately, this hypothesis has not been adequately tested. There is only one reported study of chemicals found in sea hare skin. The whole skin of *A. californica* was extracted with ether and a remarkably high concentration of prepacifenol epoxide (13)

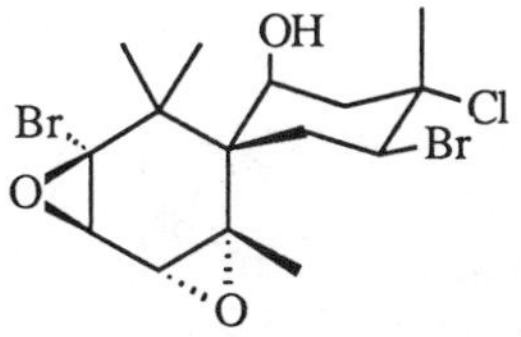

13

was recorded along with the known *Laurencia* metabolites (Stallard 1974). The high concentration of a very reactive bis-epoxide in the skin of *A. californica* may signify the selective employment of this metabolite to deter predators.

Despite all the circumstantial evidence that implicates diet-derived chemicals in the protection of sea hares, there is little direct evidence that these chemicals are effective feeding inhibitors. In the only study to specifically test sea hare metabolites, both brasilenyne (14) and *cis*-dihydrorhodophytin (15)

14   15

from *A. brasiliana* were shown to inhibit fish feeding (Kinnel et al. 1979). Several algal metabolites also found in sea hares have been investigated as part of a study of chemical defense against coral reef herbivores (Hay et al. 1987). In that study, various chemicals were coated on *Thallasia* blades at a concentration of ≈1% by weight, and the consumption of treated *Thallasia* blades by reef fishes was compared with the consumption of untreated blades over two to three hours. Elatol (7) was twice as effective as either isolaurinterol (16) or pachydictyol-A (17),

16   17

but all three compounds significantly inhibited fish feeding. On the other hand, aplysin (11), which is formed from the algal metabolite laurinterol (10) in the digestive gland of several sea hares, was quite ineffective in this assay. If, as is likely, laurinterol and isolaurinterol are equally effective as fish feeding inhibitors due to the presence of the acidic *p*-bromophenoxy group, it appears that the conversion of laurinterol into aplysin inactivates a diet-derived feeding inhibitor.

There is an obvious need for more detailed study of the various chemical mechanisms that might protect sea hares from predation. Sea hares that consume red algae should be compared and contrasted with those that eat brown algae. Particular attention should be given to sea hares that appear to eat only "inoffensive" algae such as *Ulva* spp. because these species seem to be equally free of predators. There is also a need to study the chemistry of sea hares egg masses, which are almost certainly chemically defended against predation. The egg masses do not appear to contain the same natural products found in the adult sea hares (pers. observ.), but they have been reported to contain antimicrobial macromolecules that can protect the eggs from bacterial decay (Kamiya et al. 1984).

Ascoglossans

The ascoglossans (sacoglossans) that have been studied by natural product chemists are those in the family Elysiidae that lack an external shell. These ascoglossans are herbivorous, and a number contain functional chloroplasts, obtained from siphonous marine algae, that are distributed within their diffuse digestive tracts. The chloroplasts are capable of converting bicarbonate into a variety of carbohydrates, which can be used for the synthesis of secondary metabolites. In many instances, the compounds isolated from ascoglossans are completely unrelated to metabolites from known dietary sources, and there is good evidence that these compounds are produced by the ascoglossan using organic precursors produced by the chloroplasts. This is a remarkably efficient method of obtaining a chemical defense mechanism.

There are relatively few reports of deterrent effects of ascoglossan metabolites. Lewin (1970) reported that *Oxynoe panamensis* secreted an ichthyotoxic mucus that was astringent to the tongue. Incubation of the animals with $^{14}CO_2$ produced a labeled mucus, but it is not known whether the label was restricted to the mucopolysaccharide. Tests performed on the mucus indicated that the toxin is a small molecule that could either be synthesized by the animal or obtained from its diet of *Caulerpa sertuloides*. Although Doty and Aguilar-Santos (1970) detected the *Caulerpa* metabolites caulerpin (18) and caulerpicin (19)

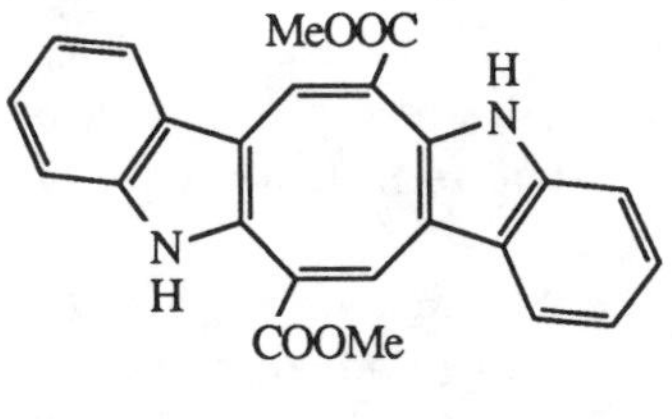

MeOOC

$$CH_3(CH_2)_{12}CH=CHCHCH-NHCO(CH_2)_nCH_3$$

with OH on the CHCH carbon and CH$_2$OH below.

18

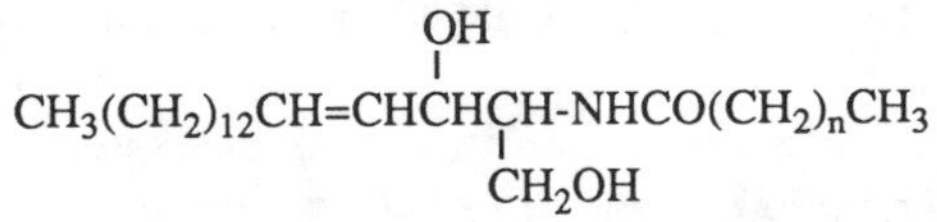

19 n = 12,14,20,22

in *O. panamensis* with thin-layer chromatography, it is not clear that these compounds were responsible for the ichthyotoxicity. Jensen (1984) reported that *Mourgona germinae*, an ascoglossan that feeds exclusively on *Cymopolia barbata*, produces a toxic mucus. It seems likely that the mucus contains brominated metabolites like cymopol (20)

20

that inhibit predation on *C. barbata* (Hay et al. 1987). A much clearer example of ingested metabolites used as antipredator defenses has been presented by Paul and Van Alstyne (1988). The ascoglossan *Elysia halimedae* obtains halimedatetraacetate (21) from *Halimeda macroloba*, reduces the aldehyde (21) into the corresponding alcohol (22),

21 R = CHO
22 R = CH₂OH

and stores the alcohol for its own defense. The alcohol is found in very high concentrations in the whole animal, in its mucus, and in egg masses. In field assays with carnivorous and herbivorous fishes on Guam, the alcohol was a significant feeding deterrent at natural concentrations.

By way of contrast, the ascoglossans *Tridachiella diomedea*, *Tridachia crispata*, *Placobranchus ocellatus*, and *Elysia chlorotica* all produce polypropionate metabolites, such as 9,10-deoxytridachione (23) (Ireland and Scheuer 1979; Ireland and Faulkner 1981; Ksebati and Schmitz 1985; Dawe and Wright 1986).

23

Ireland and Scheuer (1979) demonstrated that *P. ocellatus* can synthesize the polypropionate metabolites from bicarbonate, presumably utilizing organic compounds produced by the functional chloroplasts. Although Ireland and Scheuer proposed that the polypropionates serve only to absorb sunlight, we have recently found that tridachione (24)

24

deters feeding in *Gibbonsia elegans* at 5 μg/mg of food pellet (J. R. Pawlik, pers. comm., 1987). *Tridachiella diomedea* contains about 1 mg of tridachione per animal (Ireland and Faulkner 1981), which should be more than sufficient to deter a predatory fish.

A recent study of the ascoglossan *Cyerce nigricans* provided some confusing results: the propionate-derived pyrones isolated from this animal lacked the potent ichthyodeterrent properties of the whole animal extract (Roussis et al. 1990).

Nudibranchs

Of all the shell-less marine molluscs, the nudibranchs are the most delicate and often the most brightly colored. The coloration may serve as camouflage, as in the case of *Rostanga* species that are the same color as the sponges on which they feed (Anderson 1971), or perhaps as a warning of distastefulness (see Harris 1973; Karuso 1987). The delicate tissues of the nudibranchs appear to be an ideal food for fish, crabs, and other predators, yet reports of predation are virtually nonexistent. The secret to the apparent invulnerability of nudibranchs lies in their evolutionary history. Faulkner and Ghiselin (1983) have proposed that the nudibranchs evolved from shelled molluscs by a process that involved the gradual loss of the shell and detorsion of the visceral mass. The ancestral mollusc must have developed a diet-derived chemical (or nematocyst-based) defense mechanism prior to the diminution and eventual loss of the shell. For dorid nudibranchs, one may imagine the following sequence of events. The ancestral mollusc must first have overcome the chemical defense mechanism used by its prey, the ancestral sponge, to deter generalist predators. This presumably involved the excretion (or sequestering) rather than catabolism of the defensive compounds and required the development of a mechanism to sort the sponge's chemical constituents into those that could be catabolized and those that were either toxic or noxious. The defensive chemicals were then transferred to a location where they could be excreted in response to an attack. Only when the chemical defense mechanism was in place and fully operative could the ancestral mollusc lose its shell. It is unlikely that we will ever determine the exact sequence of events that occurred during the evolution of today's nudibranch from an ancestral shelled mollusc, nor is this knowledge particularly relevant to current chemical ecology. It is, however, important to determine which dietary constituents act as feeding inhibitors to representative predators and to discover how these compounds are stored and used by nudibranchs. Since the chemical constituents of nudibranchs have been reviewed extensively elsewhere (Faulkner 1984b, 1986, 1987, 1988a, 1990, 1991; Karuso 1987), this section concentrates on the biological activity of the metabolites.

The efficacy of the chemical defense mechanism depends on both the deterrent value of the chemicals and the ability of the nudibranch to mobilize them. The hypothesis that nudibranchs concentrate chemicals from their diet because the chemicals can then be employed defensively seems quite reasonable but needs considerable testing before it can be accepted. First, it must be shown that the chemicals can deter potential predators. Second, an attack by a predator must elicit a behavioral response that results in the predator being repelled by a chemical exudation. In the case of nudibranchs, it is very difficult to demonstrate the behavioral response because there are few reports of predation on nudibranchs. How does the investigator select a potential predator when only specialist predators are known? How should the investigator simulate an attack on a nudibranch to examine the behavioral response and measure the exudation of chemicals? To answer these questions and thereby demonstrate the existence of chemical defense mechanisms, researchers have taken a number of different approaches. While one might argue about the relative merits of individual experiments designed to evaluate deterrence, the sum of these experiments provides convincing support for the chemical defense hypothesis.

The key components of the nudibranch's defensive apparatus are the repugnatorial glands (Crozier 1917), a row of large glands situated on the mantle. These glands are separate from the mucus-producing glands and can be stimulated by irritation to exude an oily secretion. T. E. Thompson (1960) concluded that the position of these glands indicated a defensive role. Furthermore, in nudibranchs such as *Pleurobranchus* and *Philine* species, which produce acid secretions, Thompson (1983) clearly demonstrated that sulfuric acid is stored in surface acid cells or multicellular acid glands, respectively. Thompson detected sulfuric acid in the glands and in the epithelial exudate with the rhodizonate spot test or by histochemical detection of barium sulfate, but he offered no satisfactory explanation of the observation that the paper chromatogram of the acidic extract of *Philine aperta* differed from that of pure sulfuric acid. There have been no reports of histochemical studies to localize metabolites in the repugnatorial glands of nudibranchs that concentrate dietary metabolites. Defensive metabolites have been located in dorsal mantle tissues that contain the repugnatorial glands, and in some cases the chemicals have been detected in the exudate. Furthermore, a behavioral response can maximize the efficacy of chemical exudation from the dorsal mantle area. When molested, many nudibranchs can retract the rhinophores and gill tissue to expose only the dorsal mantle from which the distasteful chemicals are secreted.

One of the current debates among those studying chemical defense mechanisms is how best to determine a chemical's deterrent effect. Field bioassays using natural predators are usually preferred by ecologists, but these experiments generally require large quantities of the chemicals being tested. This is a distinct disadvantage when one is dealing with the small quantities that are usually available from nudibranchs. The alternative aquarium bioassays, which measure the palatability of chemically treated food pellets perceived by individual test animals, are critically dependent on the choice of predator. Furthermore, the screening of individual compounds rather than the precise mixture of compounds found in a nudibranch exudate might be considered a flaw in both bioassays because each compound may deter a different suite of predators. Although they are commonly performed by chemists, assays for ichthyotoxicity are probably the least valuable for determining the defensive role of a chemical because it is difficult to interrelate deterrence and toxicity. It is certainly not advantageous for an individual nudibranch to be toxic if it cannot deter predation.

It is best to take a chronological approach to a review of the palatability experiments that have contributed to our understanding of nudibranch chemical defenses. For many years, T. E. Thompson's (1969) research on acid secretions was the clearest example of the defensive use of an unpalatable chemical. The first study involving natural products was by Burreson et al. (1975), who reinvestigated a report by Johannes (1963) that the mucus of *Phyllidia varicosa* was toxic to fish and crustaceans. Johannes made the interesting observation that the mucus contained a neutral, volatile material that possessed a strong and unusual smell. Burreson et al. (1975) obtained small quantities of a volatile compound by distilling the mucus of *P. varicosa* and then found that the same compound was more readily available from a sponge of the genus *Hymeniacidon* on which the nudibranch fed. The volatile component was identified as 9-isocyanopupukeanane (25). Although there is strong circumstantial evidence that this was indeed the active material responsible for the toxicity, the compound was not assayed and was not sufficiently well characterized chemically to support future studies. The subsequent isolation of 2-isocyanopupukeanane (26)

25 R^1 = NC R^2 = H
26 R^1 = H R^2 = NC

from both sponge and nudibranch (Hagedone et al. 1979) led to a confusing situation, for it is not known whether the toxicity is due to isonitrile 25, isonitrile 26, or a combination of the two. It may be important to know whether individual compounds or classes of compounds (i.e., all isonitriles) are responsible for toxicity or feeding inhibition. Cimino et al. (1982) subsequently found axisonitrile-1 (27),

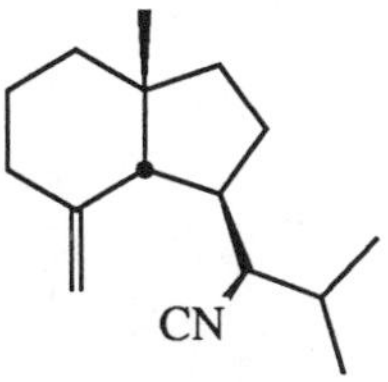

27

a metabolite of the sponge *Axinella cannibina* (Cafieri et al. 1973), to be the major constituent of the Mediterranean nudibranch *Phyllidia pulitzeri*. Axisonitrile-1 was ineffective as a fish feeding inhibitor but was active in a fish toxicity assay at a minimum concentration of 8 ppm.

A study of *Cadlina luteomarginata* from La Jolla, California, clearly showed that the secondary metabolites were of dietary origin (Thompson et al. 1982). The isolated compounds consisted of sesquiterpene and sesterterpene furans, such as furodysinin (28), idiadione (29), and pallescensin-A (30),

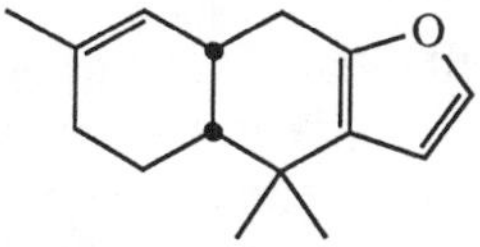

28

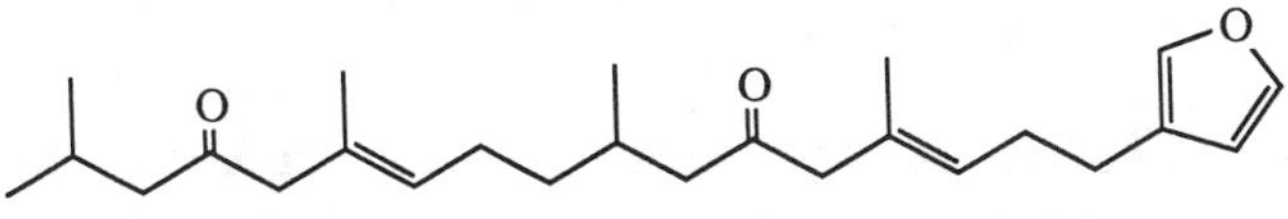

29

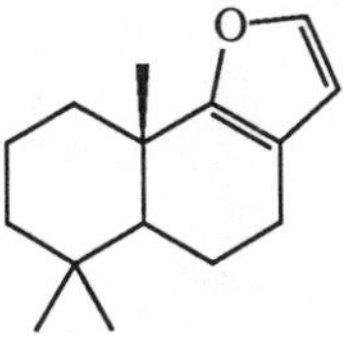

30

and several groups of sesquiterpene isonitriles, isothiocyanates, and formamide (i.e., 31–39) (Walker 1981).

31 R = NC
32 R = NCS
33 R = NHCHO

34 R = NC
35 R = NCS
36 R = NHCHO

37 R = NC
38 R = NCS
39 R = NHCHO

The terpenoids were found only in the dorsal mantle. The compounds could be traced to sponges found in the vicinity of the nudibranchs, but it was apparent that the nudibranch had exercised some control over the process of accumulating metabolites, sometimes storing only one or two compounds from an array of similar metabolites produced by the sponge. Furodysinin (28), idiadione (29), pallescensin-A (30), a "natural" mixture of isonitriles (31, 34, and 37), and a "natural" mixture of isothiocyanates (32, 34, and 38) were tested in a goldfish toxicity assay and in feeding inhibition assays using goldfish (*Carassius auratus*) and woolly sculpin (*Clinocottus analis*). All compounds were equally toxic to goldfish at 100 μg/ml but not at 10μg/ml. The isonitrile mixture was the most effective goldfish feeding inhibitor at 10 μg/mg, with furodysinin and the isothiocyanates active at 100 μg/mg. Against the sculpin, furodysinin, idiadione, and pallescensin-A were the more active feeding inhibitors at 10 μg/mg. These data together with data for the average concentrations of the metabolites in the dorsal mantle tissue indicated that *C. luteomarginata* could exude more than enough of the active chemicals to deter selected fish.

In a parallel study of *C. luteomarginata* from British Columbia, a different array of metabolites was isolated, with only one compound, furodysinin, in common with those from the La Jolla specimens (Hellou et al. 1982). This is not unexpected; the metabolites are of dietary origin and the array of sponges available to *C. luteomarginata* differs in the two locations (Faulkner et al. 1990). The major metabolite of the northern collection of *C. luteomarginata* was albicanyl acetate (40)

40

a relatively simple compound that inhibited feeding in goldfish at 5 μg/mg and was among the more potent goldfish feeding inhibitors.

Although many dorid nudibranchs concentrate furans from sponges, it is not clear whether all furans should be considered defensive chemicals. Schulte et al. (1980) reported the isolation of nakafuran-8 (41) and nakafuran-9 (42)

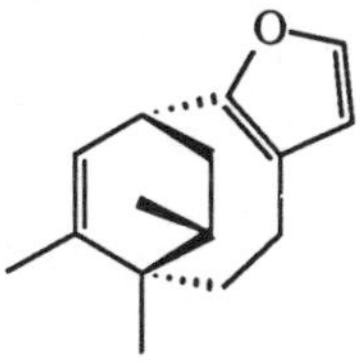

41

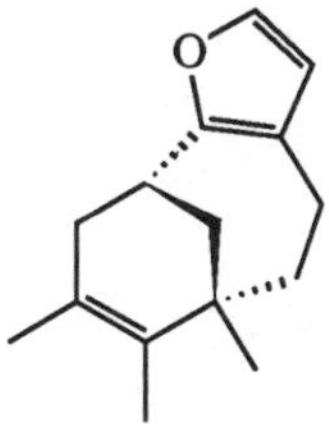

42

from the nudibranchs *Hypselodoris godeffroyana* and *Chromodoris maridadilus* and the sponge *Dysidea fragilis*. They reported that both chemicals inhibited feeding in fishes of the genus *Chaetodon* but gave no quantitative data or experimental details. Cimino et al. (1982) reported that longifolin (43),

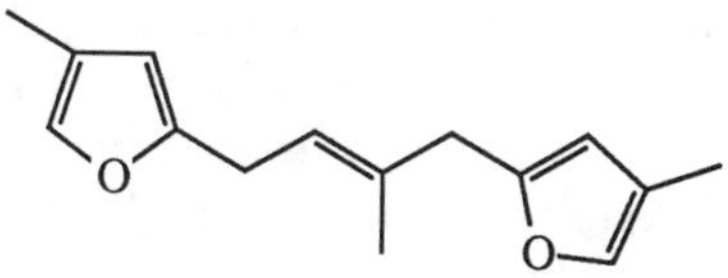

43

the major metabolite of *Glossodoris valenciennesi*, inhibited fish feeding, but only at a relatively high concentration of 300 μg/cm^2 of food flake. Carté et al. (1986b) found that furodysinin (28) was less effective as a fish feeding inhibitor than were the oxidation products O-methylfurodysinin lactone (44) and furodysinin hydroperoxide (45), which are produced from furodysinin by *Chromodoris funerea*.

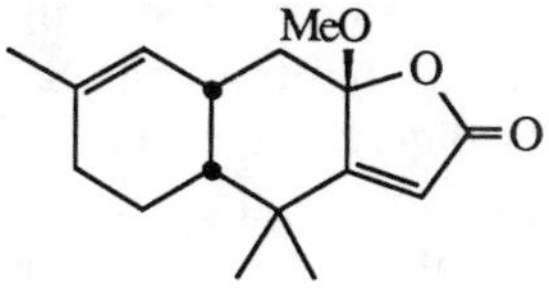

44

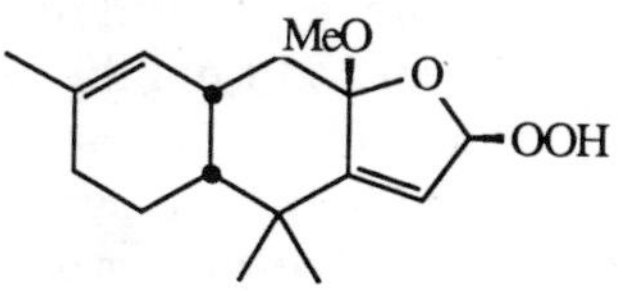

45

When tested against the spotted kelpfish (*Gibbonsia elegans*), furody-sinin caused rejection of food at a concentration of 50 μg/mg while O-methylfurodysinin lactone and furodysinin hydroperoxide were effective at 10 μg/mg and 1–5 μg/mg, respectively. Another example of a nudibranch modifying dietary chemicals to produce a more effective feeding inhibitor is *Aldisa sanguinea cooperi*, which modifies sterols to produce 3-oxo-chol-4-ene-24-oic acid (46),

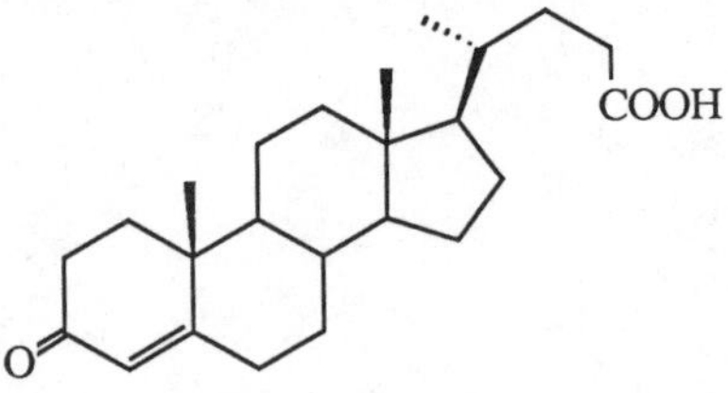

46

a compound that inhibited feeding by goldfish at a concentration of 15 μg/ml (Ayer and Andersen 1982).

The chemical defense mechanism of the porostome nudibranch *Dendrodoris limbata* has been studied in detail. Unlike most nudibranchs, which obtain chemicals from dietary sources, *D. limbata* can synthesize polygodial (47) from a labeled mevalonate precursor (Cimino et al. 1983). Polygodial, previously described as an insect antifeedant from East African Warburgia plants (Kubo et al. 1976), inhibited fish feeding at a minimum concentration of 30 μg/cm^2 on food flakes. The acyl sesquiterpene mixture (48) that might be considered either a precursor to or a degradation product of polygodial (47)

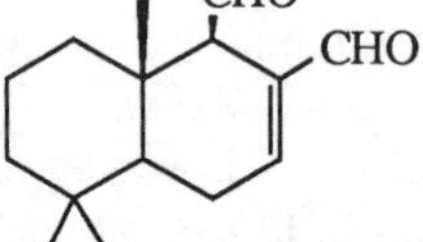

47

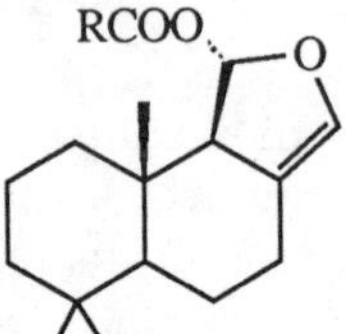

48

did not inhibit fish feeding. It is significant that the active compound (47) was isolated from the border of the dorsal mantle while the inactive mixture of esters (48) was found in the digestive gland (Avila et al. 1991), since polygodial was shown to be toxic to the nudibranch when injected into the digestive gland (Cimino et al. 1985a). Polygodial, olepupuane (49), and 6-acetoxyolepupuane (50)

49 R = H
50 R = OAc

were all subsequently isolated from several species of *Dendrodoris*, *Doriopsilla*, and other porostome nudibranchs (Okuda et al. 1983; Cimino et al. 1985b). In *Dendrodoris grandiflora*, polygodial and 6-acetoxyolepupuane, which inhibits goldfish feeding at a concentration of 40 $\mu g/cm^2$ on food flakes, were found in the mantle tissue, while a variety of furanoid metabolites of sponge origin were obtained from the digestive glands (Cimino et al. 1985b). Olepupuane, which was isolated from *Dendrodoris nigra*, *D. tuberculosa*, *D. krebsii*, *Doriopsilla albopunctata*, and *D. janaina*, inhibited feeding by the Pacific damselfish (*Dascyllus aruanus*) at a minimum concentration of 15–20 $\mu g/mg$ of food pellet.

Dendrodoris limbata, *Archidoris montereyensis*, and *A. odhneri* are among the few nudibranchs known to produce secondary metabolites. Gustafson and Andersen (1985) demonstrated that labeled mevalonate was incorporated into the diterpenoic acid glyceride (51) and sesquiterpenoic acid glyceride (52) by *A. montereyensis* and into farnesic acid glyceride (53)

51

52

53

by *A. odhneri*. In feeding inhibition assays using the sculpin *Oligocottus maculosus*, only the sesquiterpenoic acid glyceride was active at a concentration of 18 μg/mg of food pellet, while diterpenoic acid glyceride and farnesic acid glyceride were inactive at 160 μg/mg and 111 μg/mg, respectively. The glyceryl ether (54), which was also found in *A. montereyensis*, inhibited sculpin feeding at 18 μg/mg. Verrucosin-A (55) and verrucosin-B (56),

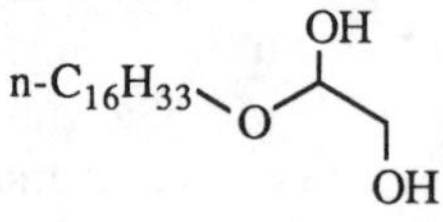

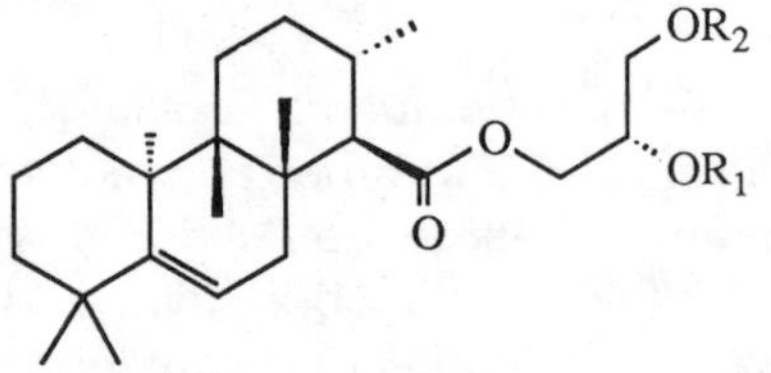

55 $R_1 = H$ $R_2 = Ac$
56 $R_1 = Ac$ $R_2 = H$

54

which are closely related structurally to the diterpenoic acid glyceride (51), were isolated from *Doris verrucosa* (Cimino et al. 1988b). The verrucosins were toxic to the mosquito fish *Gambusia affinis* at concentrations of 1.0 and 0.1 μg/ml, but farnesic acid glyceride was nontoxic.

In a study of the chemical defense of the Red Sea nudibranch *Glossodoris quadricolor*, Mebs (1985) reported that the nudibranch contained the ichthyotoxic metabolite latrunculin-B (57),

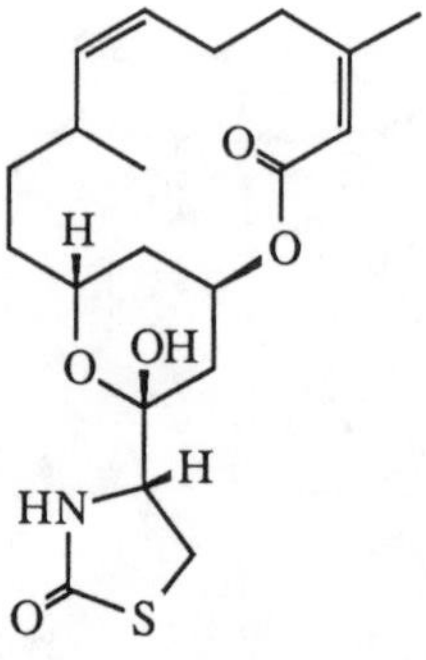

57

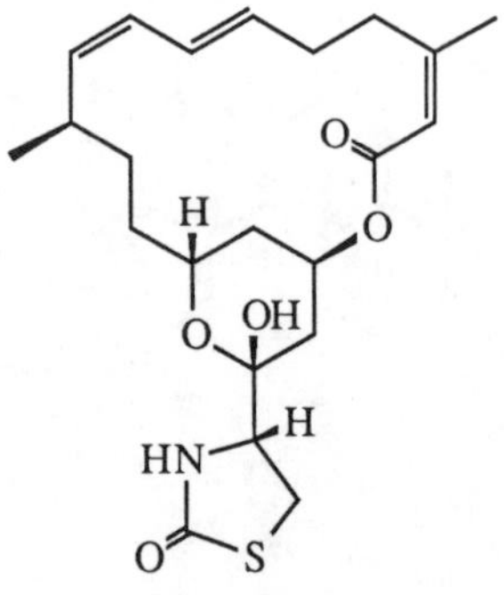

58

which it had obtained from the sponge *Latrunculia magnifica*. Latrunculin-B was also found in the mucus of *G. quadricolor*. Neèman et al. (1975) had previously demonstrated that the exudate of *L. magnifica*, from which latrunculin-A (58) and latrunculin-B (57) were isolated (Groweiss et al. 1983), caused an escape response in reef fishes. Latrunculin-A was subsequently reported as the defensive allomone of *Chromodoris elisabethae* (Okuda and Scheuer 1985) and *C. lochi* (Kakou et al. 1987). Cimino et al. (1982) reported that *Glossodoris tricolor* contained furoscalarol (59) and deoxoscalarin (60),

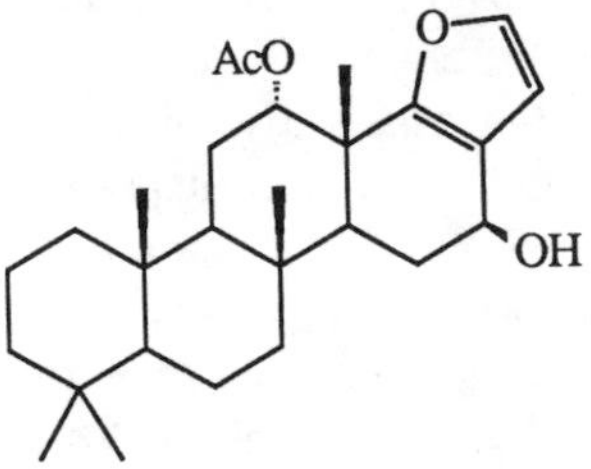

59

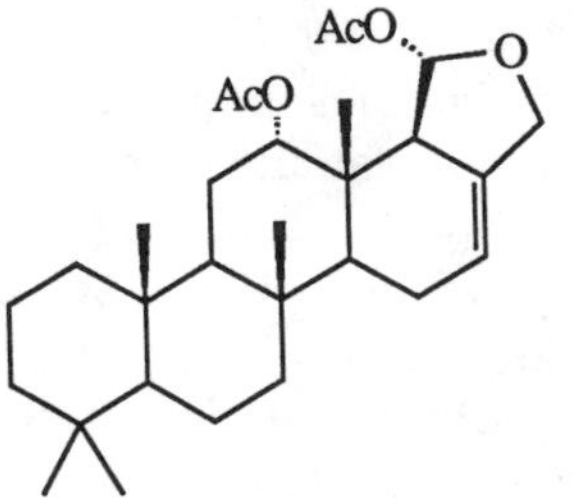

60

metabolites of the sponge *Cacospongia mollior*. Both chemicals were active as feeding inhibitors inasmuch as, after eating a small flake loaded with 250–300 $\mu g/cm^2$ of compound, the fish refused to eat even untreated food for periods ranging from a few hours to several days.

The complex relationship between the carnivorous nembrothid nudibranch *Roboastra tigris* and its preferred prey, the nembrothid nudibranchs *Tambja abdere* and *T. eliora*, centers on a chemical defense mechanism. Tambjamines A–D (61–64)

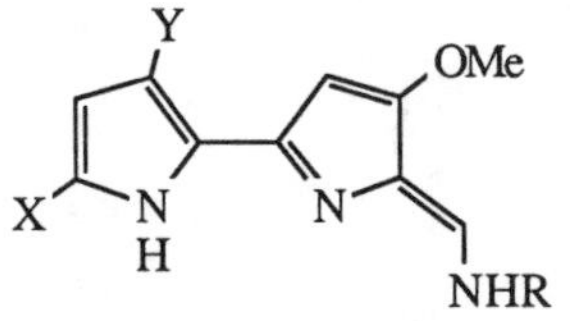

61 X = H Y = H R = H
62 X = Br Y = H R = H
63 X = H Y = H R = *i*-Bu
64 X = H Y = Br R = *i*-Bu
65 X = H Y = H R = Et
66 X = H Y = H R = CH₂CH₂Ph

are metabolites of the bryozoan *Sessibugula translucens* that are accumulated by the two *Tambja* species (Carté and Faulkner 1983). When attacked by *R. tigris*, *T. abdere*, the larger of the two *Tambja* species, exudes a yellow mucus from goblet cells in the skin; that mucus often, but not always, causes *R. tigris* to break off its attack (Carté and Faulkner 1986a). The smaller nudibranch, *T. eliora*, does not produce a defensive secretion but often escapes from *Roboastra* by using a vigorous writhing motion. Once an attack has been broken off, the *Tambja* swims away and escapes from the *R. tigris*, which must locate its prey by using contact chemoreception to follow a slime trail. It is possible that *R. tigris* is able to detect the tambjamines incorporated into the slime trail at a total concentration of around 0.36 μg/cm. The difference in the behavioral responses of the two *Tambja* species may depend on the amount of the tambjamines stored by each. When *T. abdere* was subjected to a simulated attack, it exuded 3 mg of a mixture of tambjamines, an amount known to be on the borderline of that required to deter *R. tigris*. Each *T. eliora*, however, contained less than 2 mg tambjamines, a quantity unlikely to be effective against *R. tigris*. Since the tambjamines are very effective deterrents (they inhibit feeding of the spotted kelpfish [*Gibbonsia elegans*] at concentrations of 5–10 μg/mg of food pellet), it seems quite reasonable that *T. eliora* would reserve its chemical defense mechanism to deter generalist predators and use a physical escape strategy to avoid the specialist predator that is somewhat immune to the chemical defense mechanism. A series of Y-maze experiments demonstrated that *T. eliora* can locate its food, *Sessibugula translucens*, by detecting the low concentrations of tambjamines released by the bryozoan. More remarkably, *T. eliora* could perceive a difference of two orders of magnitude in the tambjamine concentrations, for it was attracted by seawater containing tambjamines A (61) and B (62) at 10^{-10} M, but it was repelled by concentrations of 10^{-8} M and higher. Tambjamines C (63), E (65), and F (66) and the tetrapyrrole (67)

67

have recently been isolated from the ascidian *Atapozoa* sp.; these compounds are employed in the chemical defense mechanism of several species of *Nembrotha* that selectively prey on the ascidian (Lindquist 1989; Paul et al. 1990). In field assays performed in Guam, strips of dried squid that had been coated with tambjamine C at 0.4% dry weight, tambjamine F at 0.6% dry weight, or the tetrapyrrole at 0.1% dry weight were avoided by reef fishes to a significant extent.

The Indo-Pacific nudibranch *Hexabranchus sanguineus* is the largest of the dorid nudibranchs, and it accumulates significant quantities of tris-oxazole macrolides from the sponges that make up its diet

(Pawlik et al. 1988). In fact, specimens of *H. sanguineus* obtained from the Philippines would not eat fresh sponges from California but would eat frozen and thawed sponges of the genus *Halichondria* that produce the tris-oxazole macrolides. Both the reef fish *Thalassoma lunare* and the hermit crab *Dardanus megistos* rejected pieces of the mantle tissue of *H. sanguineus* but rapidly consumed similar-sized pieces of squid tissue. The macrolides ulapualide A (68) and ulapualide B (69)

68 R = O
69 R = H, OCOCH(OMe)CH$_2$OMe

and kabiramide C (70) were first isolated from the egg masses of *H. sanguineus* (Matsunaga et al. 1986; Roesener and Scheuer 1986). A number of related macrolides, including kabiramide B (71)

70 R = Me
71 R = H

and halichondramide (72), were subsequently found in sponges of the genus *Halichondria* (Kernan and Faulkner 1987). Specimens of *Hexabranchus sanguineus* found on or in the vicinity of the halichondramide-containing sponges contained 5,6-dihydrohalichondramide (73)

as the major product, and the feeding experiment outlined above demonstrated that the nudibranch had converted halichondramide into 5,6-dihydrohalichondramide. The reduced product (73) is significantly more effective as a fish feeding inhibitor than is halichondramide.

In assays using the Indo-Pacific reef fish *Thalassoma lunare*, 5,6-dihydrohalichondramide inhibited feeding at a concentration of 0.01% dry weight of pellet, a concentration well below those found in the dorsal mantle (0.70% dry weight), the digestive gland and gonad (0.47%), the mucus (0.1%), and the egg masses (1.7%), but above the concentrations found in the foot and the accessory reproductive organs (Pawlik et al. 1988). The tris-oxazole macrolides are all known to be very toxic to mice by intraperitoneal injection ($ED_{50} < 1$ mg/kg) and are extremely cytotoxic. The ability of *H. sanguineus* to handle such toxic chemicals with apparent impunity is quite remarkable.

The dendronotacean nudibranch *Tethys fimbria* stores prostaglandin lactones in the mantle and dorsolateral appendages (Cimino et al. 1991a). When molested, the nudibranch produces a mucous secretion and can easily shed the appendages. In the detached appendages, the prostaglandin lactones are coverted into PGE_2 and PGE_3, but the function of the prostaglandins, which are not ichthyotoxic, is unknown.

Although the studies described above report compounds that are effective feeding inhibitors, it would be wrong to assume that all compounds found in nudibranchs are feeding inhibitors or that all members of a class of similar compounds are equally effective. For example, in a recent study of *Chromodoris norrisi*, which concentrates selected diterpenes from the sponge *Aplysilla polyrhaphis*, the major metabolite isolated from the nudibranch was ineffective as a fish feeding inhibitor. The minor constituents, shahamin C (74) and polyrhaphin A (75), inhibited feeding by the rainbow wrasse (*Thalassoma lucasanum*) at a concentration of 100 μg/mg of brine shrimp, while the related compound macfarlandin E (76),

74   **75**   **76**

which made up about 90% of the diterpenes isolated, was not active (Bobzin and Faulkner 1989). Results from my laboratory and other hearsay results suggest that a significant number of nudibranch metabolites may be ineffective as feeding inhibitors or may have a spectrum of activities that depend on the test organism employed in the assays. These data are difficult to evaluate, as are statements that claim antifeedant properties but do not present experimental details.

Other Opisthobranchs

The cephalaspidean mollusc *Navanax inermis* is a carnivore that preys on other opisthobranchs. It employs chemotaxis to follow the slime trails of its prey, which may even include other *Navanax*. When attacked, *N. inermis* secretes a yellow hydrophobic substance from a "yellow gland" situated near the anus directly onto its slime trail. When another individual encounters this alarm substance it turns away and avoids the secretion, thereby discontinuing its pursuit of that particular slime trail. The major constituents of the yellow substance are navenones A–C (77–79) (Sleeper and Fenical 1977),

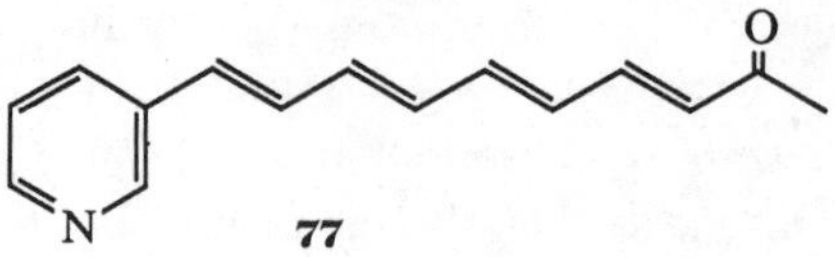

77

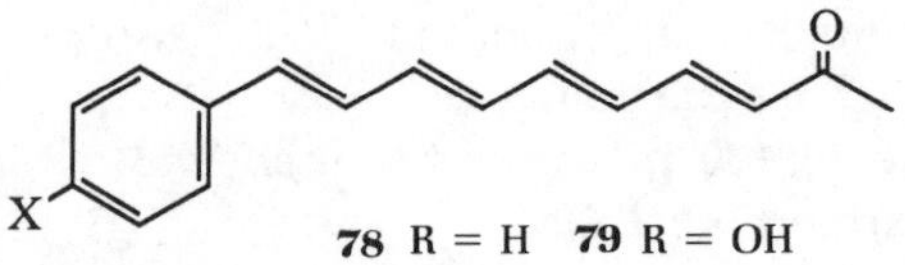

78 R = H **79** R = OH

compounds that induce the avoidance response at concentrations of
1×10^{-5} M in a seawater-slime mixture (Sleeper et al. 1980). The navenones are not of dietary origin but were synthesized by *N. inermis*
from labeled sodium acetate (Fenical et al. 1979). Similar compounds
are used by *Haminoea navicula* to induce an alarm response in trail-
following conspecifics (Cimino et al. 1991b).

The notaspidean mollusc *Umbraculum mediterraneum* contains
two unusual diacylglycerols, umbraculumin-A (80) and umbraculu-
min-C (81),

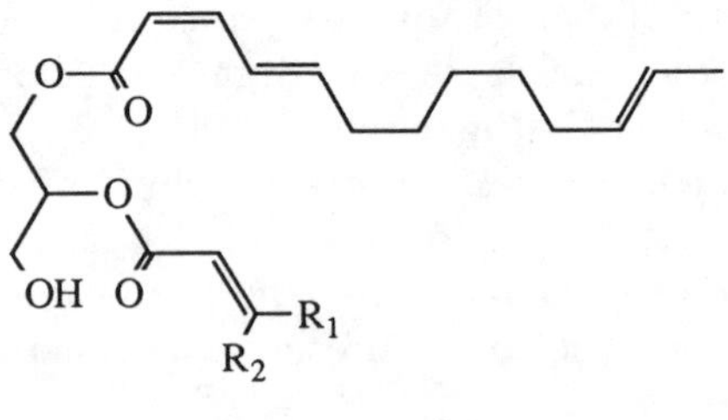

80 $R_1 = R_2 = Me$
81 $R_1 = SMe \ R_2 = H$

which are toxic to the mosquito fish *Gambusia affinis* at concentrations of 10 μg/ml and 0.1 μg/ml, respectively (Cimino et al. 1988a). Another notaspidean mollusc, *Tylodina fungina*, is always found feeding
on *Aplysina* spp. and invariably contains the same brominated metabolites as its food source (pers. observ.). Although *T. fungina* is partially
covered by a soft shell, the brominated metabolites could provide considerable protection from predators. J. E. Thompson et al. (1985) reported that aerothionin (82),

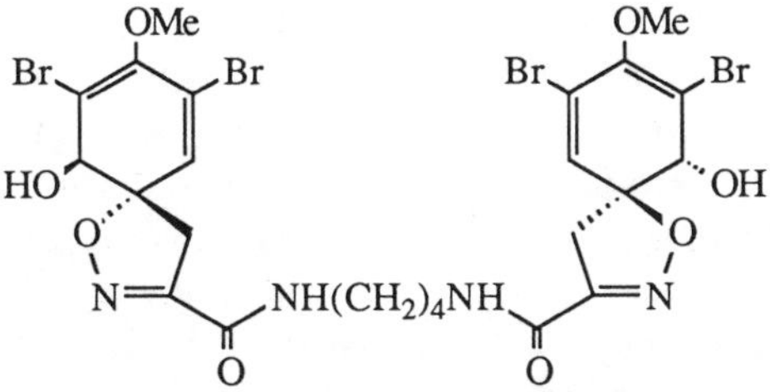

82

which has been found in some specimens of *T. fungina*, inhibited feeding in the sculpin *Clinocottus analis* at a concentration of 10 μg/ml of food pellet and was toxic to several invertebrate species.

The bubble shell (*Haminoea cymbalum*) contains kumepaloxane (83),

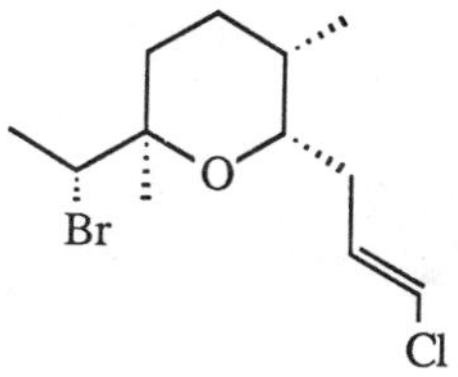

83

an unusual halogenated metabolite of unknown origin. Kumepaloxane is exuded in the mucus when the animal is disturbed and is an effective feeding deterrent against carnivorous fishes in both field and aquarium assays (Poiner et al. 1989).

Marine Pulmonates

The pulmonates are distinguished by their ability to breathe air and are principally terrestrial or freshwater molluscs. The marine pulmonates can "breathe" both above and under water and are especially adapted to living in the intertidal zone. The chemical ecology of three groups—the onchiids, the siphonariids, and the trimusculiids—has been studied in some detail. The onchiids have a leathery skin but no shell, while the siphonariids and trimusculiids have shells of various strengths. These pulmonate studies may therefore provide insight into the relative value of physical and chemical defense mechanisms.

The chemical defense mechanism of onchiids is one of the easiest to observe. When molested, these animals exude a white mucus secretion from apical pores in papillae situated around the edge of the mantle. Arey and Crozier (1921) demonstrated that the secretion of *Onchidella floridanum* acted as a deterrent to several potential predators, including fish and crabs; and Young et al. (1986) reported that the secretion of *O. borealis* repelled intertidal predatory asteroids. Studies using *Onchidella binneyi* revealed that the mucus secretion contained only onchidal (84)

84

in a mucopolysaccharide matrix (Ireland and Faulkner 1978). Recent investigations of a number of *Onchidella* species, including *O. borealis* from northern California, *O. nigricans* from New Zealand, and *O. patelloides* from Australia, demonstrated that all specimens contained onchidal. This wide distribution suggests that onchidal is not a dietary constituent but is produced by the animals (Manker 1988). Onchidal is toxic to goldfish at concentrations greater than 36 μM. It acts at the cellular level by irreversibly inhibiting acetylcholinesterase (Abramson et al. 1989).

The siphonariids are characterized by their ability to synthesize polypropionate metabolites (Manker et al. 1988). When molested, *Siphonaria* spp. produce a white mucus that contains the polypropionate metabolites. Since the *Siphonaria* spp. are physically protected by a shell, the defensive value of the polypropionates is questionable, particularly since very few compounds have been evaluated as feeding inhibitors, and only vallartanone B (85), but not vallartanone A (86),

85 R = CH$_3$
86 R = H

with which it coexists in *Siphonaria maura* from Mexico, inhibits feeding by the reef fish *Thalassoma lunare* at 100 μg/mg on krill (Manker and Faulkner 1989). The fact that the only *Siphonaria* species examined that did not contain polypropionate metabolites was *S. gigas*, an animal that possesses a very heavy shell, suggests that there might be an inverse relationship between chemical and physical protection in the siphonariids. Siphonariids invariably coexist with limpets. In field experiments in which *Siphonaria maura* and co-occurring limpets were detached from the rocks and offered to tidepool predators (blennies, sculpins, hermit crabs, etc.), both siphonariids and limpets were eaten with equal enthusiasm (Manker 1988). Similar observations were made for *S. denticulata* and *S. virgulata* in Australia. Experiments to evaluate the response of the larvae of a polychaete worm (*Phragmatopoma lapidosa californica*) to polypropionates indicated that pectinatone (87) from *S. pectinata* (revised structure from Garson et al. 1990), a mixture of denticulatin A (88) and denticulatin B (89) from *S. denticulata*, and a furanone mixture (90) from *S. thersites*

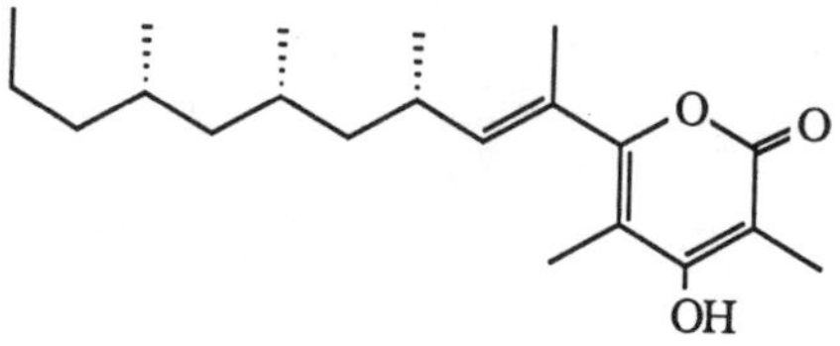

87

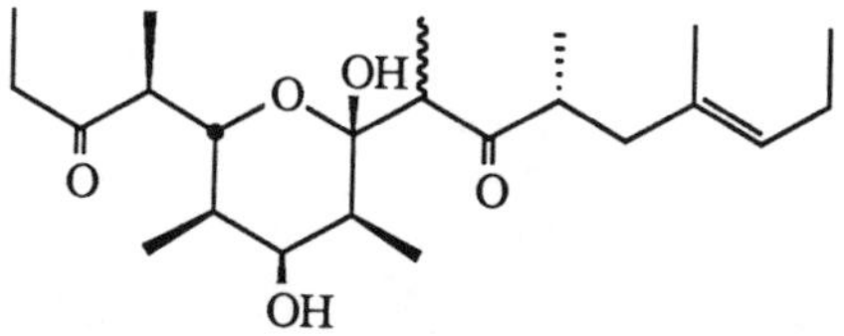

88 β - Me
89 α - Me

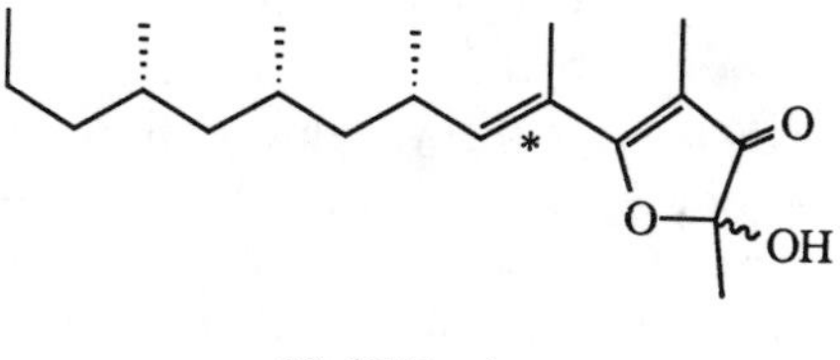

90 *E/Z* mixture

were all larvicidal when coated on sand at a load of 100 $\mu g/g$ in seawater (10 ml), while vallartanones A (86) and B (85) caused some of the larvae to settle (Manker 1988). Thus it is possible that polypropionate metabolites may function to prevent competing invertebrates from settling. In evaluating the relative importance of chemical and physical defenses to the siphonariids, one it must recognize that the shell is needed to combat the potential damage caused by wave action and dehydration. When firmly attached to a rock, the siphonariid is probably protected from most predators by its shell and would need further protection only when foraging for food or when dislodged. It is probable that the most important functions of the polypropionates have yet to be discovered and may not be related to chemical defense at all.

Trimusculiid pulmonates are completely sessile molluscs that live only in narrow crevices and deep caves, where they form large colonies hidden from most predators. They feed by producing a mucus net to trap suspended phytoplankton, which are then ingested (Walsby et al. 1973). The trimusculiids have mucous glands along the edge of the mantle, and Yonge (1957) proposed that the mucus produced by these glands might be repugnatorial. Rice (1985) demonstrated that the mucus produced by *Trimusculus reticulatus* repelled the predatory starfish *Pisaster ochraceus* and *P. giganteus*. The foot and mantle of *T. reticulatus* contained diterpenes (91 and 92),

91 R_1 = H R_2 = H
92 R_1 = Ac R_2 = OAc

but only the major metabolite (91) was found in the mucus. A solution of diterpene (91) did not repel the starfish, but it is possible that repugnatorial effects may require a tactile response. Both the mucus and the diterpene (91) were toxic to the larvae of the worm *Phragmatopoma*

lapidosa californica when coated on sand at a concentration of 100 μg
on 1 g sand in 10 ml seawater (Manker 1988). For a sessile animal like
T. reticulatus, it could be advantageous to restrict the growth of competing sessile invertebrates by inhibiting larval settling.

Limpets

Limpets are normally protected from predation by their ability to
remain firmly attached to rocks so that the shell affords complete protection. When detached from rocks, however, they are susceptible to a
number of fish and invertebrate predators (for review, see Branch
1981). Among the limpets of the California coast, *Lottia* (= *Collisella*)
limatula is unique in possessing a chemical defense mechanism that
protects it against tidepool predators (Pawlik et al. 1986). The foot of *L.
limatula* contains limatulone (93) (Albizati et al. 1985)

93

which inhibits feeding in the spotted kelpfish (*Gibbonsia elegans*) at a
concentration of 0.5 ppt in food pellets. The rationale for the presence
of a defensive metabolite is fairly complex. *L. limatula* can absorb the
force of a sharp blow caused by wave-borne rocks and debris by allow-
ing the margin of the shell to break off, thus exposing the foot to predators. Chemical protection of the foot tissue allows *L. limatula* to survive loss of the shell margin, which in turn enables it to live in wave-
battered and exposed locations where other limpets cannot survive.
The study described above is important in another respect: it is the
only study of a chemical defense mechanism in molluscs that was
performed in an unbiased manner. The study followed the classical
pattern of identifying an unpalatable organism, devising a suitable bio-
assay, and performing a bioassay-guided fractionation to identify the
active compound.

Prosobranch Molluscs

It is rare that a prosobranch mollusc does not have a heavy shell, but this is the case for certain lamellariids that feed on ascidians and have a diminished shell. In the one example that has been studied, a *Lamellaria* species contained a mixture of aromatic metabolites called lamellarins A–D (94–97) (Andersen et al. 1985).

94 X = OH   96 X = H

95 X = OMe R = Me
97 X = H   R = H

Almost identical compounds were subsequently found in the ascidian *Didemnum chartaceum* (Lindquist et al. 1988). V. J. Paul (pers. comm., 1989) has shown that a mixture of lamellarins is unpalatable to reef fishes.

The egg cowrie (*Ovula ovum*) feeds on the soft coral *Sarcophyton* sp. and converts the major coral metabolite sarcophytoxide (98) into 7,8-deoxysarcophytoxide (99).

98   99

The conversion is considered to be a detoxification mechanism since the product is much less toxic to *Gambusia affinis* than is sarcophytoxide (Coll et al. 1983).

Secondary metabolites have been reported from several other prosobranchs, but there is no suggestion that these compounds are involved in chemical defense.

Conclusions

The overall picture that has emerged from the studies outlined above is that chemical secretions produced by opisthobranch molluscs can act to reduce predation. The data are by no means perfect, for not every compound tested has been shown to inhibit predation. Yet it can be argued that the defensive chemicals need not deter every predator. Instead they should diminish predation to the extent that individuals possessing the chemicals survive and reproduce to a greater extent than those lacking defensive chemicals. The relationship between chemical and physical defenses is therefore one of balance: the mollusc must have either physical defenses, a chemical defense mechanism, or an effective combination of both. There is no reason why a mollusc should not have both a shell to protect it against physical dangers and chemicals to deter predators, but in this case it may be more difficult to demonstrate the defensive value of the chemical secretions because the mollusc may rely less on chemicals.

There is an urgent need for more research, particularly interdisciplinary studies, to answer the many questions that remain in this field. It is not clear whether opisthobranch molluscs use chemoreception or chemotaxis to detect their food, or to what extent they require a specific food for survival. Research on *Hexabranchus sanguineus* suggested a very specific dietary requirement (Pawlik et al. 1988), but studies on *Chromodoris funerea* (Kernan et al. 1988) showed that mollusc to be quite flexible in its diet. There is evidence that a nudibranch can adapt to its surroundings and concentrate the most effective feeding deterrents from the sponges that are locally available (J. E. Thompson et al. 1982; Hellou et al. 1982). If this is true, how does the nudibranch determine which chemical is of greatest defensive value? There has been no satisfactory explanation of how the opisthobranchs distinguish between dietary constituents. If the nudibranch requires an inherited enzymatic apparatus to separate specific molecules, then it is implied that the nudibranch can only store specific chemicals and will always seek out only those sponges that contain them. If, on the other hand, the nudibranch can "taste" each chemical and separate chemicals on the basis of taste, then it will be able to select the best available defensive arsenal. Whatever the case, the biochemical mechanism for sorting chemicals is unknown and should be a priority for further studies.

There remains a need for new bioassays that can take into account both palatability and the behavioral aspects of feeding. Experiments to date suggest that the apparent lack of predation on opisthobranchs is due to more than just unpalatability. When screening compounds for palatability, there needs to be more emphasis on evaluating both the crude exudation and individual compounds at the same time, using the same assay animals. Such assays will disclose potential synergisms. The function of the mucus deserves further investigation, and there is a need for a synthetic "mucus" that would allow researchers to study the chemotactic effects of pure chemicals that are normally mobilized in mucous secretions.

Researchers who wish to do further studies on chemical deterrence in opisthobranchs should seek an interdisciplinary environment that will enable specialists to combine their chemical and ecological skills. Most of all, there is a need to perform interdisciplinary studies that employ bioassay-guided fractionations in order to isolate the active rather than the major constituents of exudates.

Acknowledgments I am grateful for the contributions of my colleagues at the Scripps Institution of Oceanography who have, over the years, introduced me to the philosophical differences between chemistry and ecology. My interest in the chemical defenses of marine molluscs has been enthusiastically shared by a talented group of researchers, including Kim F. Albizati, Raymond J. Andersen, Steven C. Bobzin, Brad Carté, Mary Kay Harper, Jill Hochlowski, Chris Ireland, Michael R. Kernan, Denise C. Manker, Tadeusz F. Molinski, Joseph R. Pawlik, Janice E. Thompson, Roger P. Walker, and Stephen J. Wratten. I thank Thomas H. Carefoot, Joseph R. Pawlik, and Steven C. Pennings for critically reviewing the manuscript. Research in my laboratory has been supported primarily by grants from the National Science Foundation.

References

Abramson, S.N., Radic, Z., Manker, D., Faulkner, D.J., and Taylor, P. 1989. Onchidal: a naturally occurring irreversible inhibitor of acetylcholinesterase with a novel mechanism of action. Mol. Pharmacol. 36:349–354.
Albizati, K.F., Pawlik, J.R., and Faulkner, D.J. 1985. Limatulone, a potent defensive metabolite of the intertidal limpet *Collisella limatula*. J. Org. Chem. 50:3428–3430.
Andersen, R.J., Faulkner, D.J., He, C.-H., Van Duyne, G., and Clardy, J. 1985. Metabolites of the marine prosobranch mollusc *Lamellaria* sp. J. Am. Chem. Soc. 107:5492–5495.
Anderson, E.S. 1971. The association of the nudibranch *Rostanga pulchra* MacFarland, 1905, with the sponges *Ophlitaspongia pennata*, *Esperiopsis originalis*, and *Plocamia karykina*. Ph.D. dissertation, University of California, Santa Cruz.

Arey, L.B., and Crozier, W.J. 1921. On the natural history of onchidium. J. Exp. Zool. 32:443–502.

Avila, C., Cimino, G., Crispino, A., and Spinella, A. 1991. Drimane sesquiterpenoids in Mediterranean *Dendrodoris* nudibranchs: Anatomical distribution and biological role. Experientia 47:306–310.

Ayer, S.W., and Andersen, R.J. 1982. Steroidal antifeedants from the dorid nudibranch *Aldisa sanguinea cooperi*. Tetrahedron Lett. 23:1039–1042.

Baker, B., Ratnapala, L., Mahindaratne, M.P.D., de Silva, E.D., Tillekeratne, L.M.V., Jeong, J.W., Scheuer, P.J., and Seff, K. 1988. Lankalapuol A and B: two *cis*-eudesmanes from the sea hare *Aplysia dactylomela*. Tetrahedron 44:4695–4701.

Blankenship, J.E., Langlais, P.J., and Kittredge, J.S. 1975. Identification of a cholinomimetic compound in the digestive gland of *Aplysia californica*. Comp. Biochem. Physiol. C 51:129–137.

Bobzin, S.C., and Faulkner, D.J. 1989. Diterpenes from the marine sponge *Aplysilla polyrhaphis* and the dorid nudibranch *Chromodoris norrisi*. J. Org. Chem. 54:3902–3907.

Branch, G.M. 1981. The biology of limpets: physical factors, energy flow and ecological interactions. Oceanogr. Mar. Biol. Annu. Rev. 19:235–379.

Burreson, B.J., Clardy, J., Finer, J., and Scheuer, P.J. 1975. 9-Isocyanopupukeanane, a marine invertebrate allomone with a new sesquiterpene skeleton. J. Am. Chem. Soc. 97:4763–4764.

Cafieri, F., Fattorusso, E., Magno, S., Santacroce, C., and Sica, D. 1973. Isolation and structure of axisonitrile-1 and axisothiocyanate-1, two unusual sesquiterpenoids from the marine sponge *Axinella cannabina*. Tetrahedron 29:4259–4262.

Carefoot, T.H. 1987. *Aplysia*: its biology and ecology. Oceanogr. Mar. Biol. Annu. Rev. 25:167–284.

Carew, T.J., and Kupfermann, I. 1974. The influence of different natural environments on habituation in *Aplysia californica*. Behav. Biol. 12:339–345.

Carté, B., and Faulkner, D.J. 1983. Defensive metabolites from three nembrothid nudibranchs. J. Org. Chem. 48:2314–2318.

Carté, B., and Faulkner, D.J. 1986a. Role of secondary metabolites in feeding associations between a predatory nudibranch, two grazing nudibranchs, and a bryozoan. J. Chem. Ecol. 12:795–804.

Carté, B., Kernan, M.R., Barrabee, E.B., and Faulkner, D.J. 1986b. Metabolites of the nudibranch *Chromodoris funerea* and the singlet oxygen oxidation products of furodysin and furodysinin. J. Org. Chem. 51:3528–3532.

Chapman, D.J., and Fox, D.L. 1969. Bile pigment metabolism in the sea hare *Aplysia*. J. Exp. Mar. Biol. Ecol. 4:71–78.

Cimino, G., Crispino, A., Di Marzo, V., Spinella, A., and Villani, G. 1991a. A marine mollusc provides the first example of in vivo storage of prostaglandins: Prostaglandin-1,15-lactones. Experientia 47:56–60.

Cimino, G., Crispino, A., Spinella, A., and Sodano, G. 1988a. Two ichthyotoxic diacylglycerols from the opisthobranch mollusc *Umbraculum mediterraneum*. Tetrahedron Lett. 29:3613–3616.

Cimino, G., De Rosa, S., De Stefano, S., Morrone, R., and Sodano, G. 1985b. The chemical defense of nudibranch molluscs: structure, biosynthetic origin and defensive properties of terpenoids from the dorid nudibranch *Dendrodoris grandiflora*. Tetrahedron 41:1093–1100.

Cimino, G., De Rosa, S., De Stefano, S., and Sodano, G. 1982. The chemical defense of four Mediterranean nudibranchs. Comp. Biochem. Physiol. B 73:471–474.

Cimino, G., De Rosa, S., De Stefano, S., Sodano, G., and Villani, G. 1983. Dorid nudibranch elaborates its own chemical defense. Science 219:1237–1238.

Cimino, G., De Rosa, S., De Stefano, S., and Sodano, G. 1985a. Observations on the toxicity and metabolic relationships of polygodial, the chemical defense of the nudibranch *Dendrodoris limbata*. Experientia 41:1335–1336.

Cimino, G., Gavagnin, M., Sodano, G., Puliti, R., Mattia, C. A., and Mazzarella, L. 1988b. Verrucosin-A and -B, ichthyotoxic diterpenoic acid glycerides with a new carbon skeleton from the dorid nudibranch *Doris verrucosa*. Tetrahedron 44:2301–2310.

Cimino, G., Gavagnin, M., Sodano, G., Spinella, A., Strazzullo, G., Schmitz, F.J., and Gopichand, Y. 1987. Revised structure of bursatellin. J. Org. Chem. 52:2301–2303.

Cimino, G., Passeggio, A., Sadano, G., Spinella, A., and Villani, G. 1991b. Alarm pheromones from the mediterranean opisthobranch *Haminoea navicula*. Experientia 47:61–63.

Coll, J.C., Tapiolas, D.M., Bowden, B.F., Webb, L., and Marsh, H. 1983. Transformation of soft coral (Coelenterata: Octocorallia) terpenes by *Ovula ovum* (Mollusca: Prosobranchia). Mar. Biol. 74:35–40.

Crozier, W.J. 1917. The nature of the conical bodies on the mantle of certain nudibranchs. Nautilus 30:103–106.

Danise, B., Minale, L., Riccio, R., Amico, V., Oriente, G., Piattelli, M., Tringali, C., Fattorusso, E., Magno, S., and Mayol, L. 1977. Further perhydroazulene diterpenes from marine organisms. Experientia 33:413–415.

Dawe, R.D., and Wright, J.L.C. 1986. The major polypropionate metabolites from the sacoglossan mollusc *Elysia chlorotica*. Tetrahedron Lett. 27:2559–2562.

de Silva, E.D., Schwartz, R.E., Scheuer, P.J., and Shoolery, J. N. 1983. Srilankenyne, a new metabolite from the sea hare *Aplysia oculifera*. J. Org. Chem. 48:395–396.

Dieter, R.K., Kinnel, R., Meinwald, J., and Eisner, T. 1979. Brasudol and isobrasudol: two bromosesquiterpenes from a sea hare (*Aplysia brasiliana*). Tetrahedron Lett., pp. 1645–1648.

DiMatteo, T. 1981. The inking behavior of *Aplysia dactylomela* (Gastropoda: Opisthobranchia): evidence for distastefulness. Mar. Behav. Physiol. 7:285–290.

DiMatteo, T. 1982. The ink of *Aplysia dactylomela* (Rang, 1828) (Gastropoda: Opisthobranchia) and its role as a defensive mechanism. J. Exp. Mar. Biol. Ecol. 57:169–180.

Doty, M.S., and Aguilar-Santos, G. 1970. Transfer of toxic algal substances in marine food chains. Science 24:351–355.

Faulkner, D.J. 1984a. Marine natural products: metabolites of marine algae and herbivorous marine molluscs. Nat. Prod. Rep. 1:251–280.

Faulkner, D.J. 1984b. Marine natural products: metabolites of marine invertebrates. Nat. Prod. Rep. 1:551–598.

Faulkner, D.J. 1986. Marine natural products. Nat. Prod. Rep. 3:1–33.

Faulkner, D.J. 1987. Marine natural products. Nat. Prod. Rep. 4:539–576.

Faulkner, D.J. 1988a. Marine natural products. Nat. Prod. Rep. 5:613–663.

Faulkner, D.J. 1988b. Feeding deterrents in molluscs. *In* Biomedical importance of marine organisms, ed. D.G. Fautin, pp. 29–36. San Francisco: California Academy of Sciences.

Faulkner, D.J. 1990. Marine natural products. Nat. Prod. Rep. 7:269–309.

Faulkner, D.J. 1991. Marine natural products. Nat. Prod. Rep. 8:97–147.

Faulkner, D.J., and Ghiselin, M.T. 1983. Chemical defense and evolutionary ecology of dorid nudibranchs and some other opisthobranch gastropods. Mar. Ecol. Prog. Ser. 13:295–301.

Faulkner, D.J., Molinski, T.F., Dumdeí, E.J., de Silva, E.D., Andersen, R.J. 1990. Geographical variation in defensive chemicals from Pacific coast dorid nudibranchs and some related molluscs. Comp. Biochem. Physiol. C 97:233–240.

Faulkner, D.J., and Stallard, M.O. 1973. 7-Chloro-3,7-dimethyl-1,4,6-tribromo-1-octen-3-ol, a novel monoterpene alcohol from *Aplysia californica.* Tetrahedron Lett., pp. 1171–1174.

Faulkner, D.J., Stallard, M.O., Fayos, J., and Clardy, J. 1973. (3R,4S 7S)-*trans,trans*-3,7-Dimethyl-1,8,8-tribromo-3,4,7- trichloro-1,5-octadiene, a novel monoterpene from the sea hare, *Aplysia californica.* J. Am. Chem. Soc. 95:3413–3414.

Fenical, W., Sleeper, H.L., Paul, V.J., Stallard, M.O., and Sun, H. H. 1979. Defensive chemistry of *Navanax* and related opisthobranch molluscs. Pure Appl. Chem. 51:1865–1874.

Flury, F. 1915. Über das Aplysiengift. Arch. Exp. Pathol. Pharmakol. 79:250–263.

Garson, M.J., Small, C.J., O'Hagan, D., Skelton, B.W., Thinapong, P., and White, A.H. 1990. Sereochemical correlations of polypropionate metabolites from marine pulmonates: revision of the relative stereochemistry of pectinatone by X-ray structure analysis. J. Chem. Soc. Perkin Trans. I, pp. 805–807.

Gerwick, W.H., and Whatley, G. 1989. *Aplysia* sea hare assimilation of secondary metabolite from brown seaweed, *Stypopodium zonale.* J. Chem. Ecol. 15:677–683.

González, A.G., Cataldo, F., Fernández, J., and Norte, M. 1987. A new trioxygenated diterpene from the mollusk *Aplysia dactylomela.* J. Nat. Prod. 50:1158–1159.

González, A.G., Martín, J.D., Norte, M., Pérez, R., Weyler, V., Perales, A., and Fayos, J. 1983a. New halogenated constituents of the digestive gland of the sea hare *Aplysia dactylomela.* Tetrahedron Lett. 24:847–848.

González, A.G., Martín, J.D., Norte, M., Pérez, R., Weyler, V., Rafii, S., and Clardy, J. 1983b. A new diterpene from *Aplysia dactylomela.* Tetrahedron Lett. 24:1075–1076.

Gopichand, Y., and Schmitz, F.J. 1980. Bursatellin: a new diol dinitrile from the sea hare *Bursatella leachii pleii.* J. Org. Chem. 45:5383–5385.

Gopichand, Y., Schmitz, F.J., Shelly, J., Rahman, A., and van der Helm, D. 1981. Marine natural products: halogenated acetylenic ethers from the sea hare *Aplysia dactylomela.* J. Org. Chem. 46:5192–5197.

Groweiss, A., Shmueli, U., and Kashman, Y. 1983. Marine toxins of *Latrunculia magnifica.* J. Org. Chem. 48:3512–3516.

Gustafson, K., and Andersen, R.J. 1985. Chemical studies of British Columbia nudibranchs. Tetrahedron 41:1101–1108.

Hagedone, M.R., Burreson, B.J., Scheuer, P.J., Finer, J.S., and Clardy, J. 1979. Defense allomones of the nudibranch *Phyllidia varicosa* Lamarck 1801. Helv. Chim. Acta 62:2484–2494.

Harris, L.G. 1973. Nudibranch associations. *In* Current topics in comparative pathobiology, vol. 2, ed. C. T. Cheng, pp. 286–291. Baltimore: Academic Press.

Hay, M.E., Fenical, W., and Gustafson, K. 1987. Chemical defense against diverse coral-reef herbivores. Ecology 68:1581–1591.

Hellou, J., Andersen, R.J., and Thompson, J.E. 1982. Terpenoids from the dorid nudibranch *Cadlina luteomarginata.* Tetrahedron 38:1875–1879.

Hollenbeak, K.H., Schmitz, F.J., Hossain, M.B., and van der Helm, D. 1979. Marine natural products: deodactol, antineoplastic sesquiterpenoid from the sea hare *Aplysia dactylomela.* Tetrahedron 35:541–545.

Ichiba, T., and Higa, T. 1986. New cuparene-derived sesquiterpenes with unprecedented oxygenation patterns from the sea hare *Aplysia dactylomela.* J. Org. Chem. 51:3364–3366.

Imperato, F., Minale, L., and Riccio, R. 1977. Constituents of the digestive gland of molluscs of the genus *Aplysia*. II. Halogenated monoterpenes from *Aplysia limacina*. Experientia 33:1273–1274.

Inouye, Y., Uchida, H., Kusumi, T., and Kakisawa, H. 1987. Structure and absolute configuration of aplysiapyranoid B, a new polyhalogenated pyranoid monoterpene from *Aplysia kurodai*. J. Chem. Soc. Chem. Commun. 346–347.

Ireland, C., and Faulkner, D.J. 1977. Diterpenes from *Dolabella californica*. J. Org. Chem. 43:3157–3162.

Ireland, C., and Faulkner, D.J. 1978. The defensive secretion of the opisthobranch mollusc *Onchidella binneyi*. Bioorg. Chem. 7:125–131.

Ireland, C., and Faulkner, D.J. 1981. The metabolites of the marine molluscs *Tridachiella diomedea* and *Tridachia crispata*. Tetrahedron 37:233–240.

Ireland, C., Faulkner, D.J., Finer, J., and Clardy, J. 1976. A novel diterpene from *Dolabella californica*. J. Am. Chem. Soc. 98:4664–4665.

Ireland, C., and Scheuer, P.J. 1979. Photosynthetic marine mollusks: In vivo ^{14}C incorporation into metabolites of the sacoglossan *Placobranchus ocellatus*. Science 205:922–923.

Ireland, C., Stallard, M.O., Faulkner, D.J., Finer, J., and Clardy, J. 1976. Some chemical constituents of the digestive gland of the sea hare *Aplysia californica*. J. Org. Chem. 41:2461–2465.

Jensen, K.R. 1984. Defensive behavior and toxicity of ascoglossan opisthobranch *Mourgona germaineae* Marcus. J. Chem Ecol. 10:476–486.

Johannes, R.E. 1963. A poison-secreting nudibranch (Mollusca: Opisthobranchia). Veliger 5:104–105.

Kakou, Y., Crews, P., and Bakus, G.J. 1987. Dendrolasin and latrunculin A from the Fijian sponge *Spongia mycofijiensis* and an associated nudibranch *Chromodoris lochi*. J. Nat. Prod. 50:482–484.

Kamiya, H., Muramoto, K., and Ogata, K. 1984. Antibacterial activity in the egg mass of a sea hare. Experientia 40:947–949.

Karuso, P. 1987. Chemical ecology of the nudibranchs. *In* Bioorganic marine chemistry, vol. 1, ed. P. J. Scheuer, pp. 31–60. Berlin: Springer-Verlag.

Katayama, A., Ina, K., Nozaki, H., and Nakayama, M. 1982. Structure elucidation of kurodainol, a novel halogenated monoterpene from sea hare (*Aplysia kurodai*). Agric. Biol. Chem. 43:859–860.

Kato, Y., and Scheuer, P.J. 1974. Aplysiatoxin and debromoaplysiatoxin, constituents of the marine mollusk *Stylocheilus longicauda* (Quoy and Gaimard, 1824). J. Am. Chem. Soc. 96:2245–2246.

Kernan, M.R., Barrabee, E.B., and Faulkner, D.J. 1988. Variation of the metabolites of *Chromodoris funerea*: comparison of specimens from a Palauan marine lake with those from adjacent waters. Comp. Biochem. Physiol. B 89:275–278.

Kernan, M.R., and Faulkner, D.J. 1987. Halichondramide, an antifungal macrolide from the sponge *Halichondria* sp. Tetrahedron Lett. 28:2809–2812.

Kigoshi, H., Imamura, Y., Yoshikawa, K., and Yamada, K. 1990. Three new cytotoxic alkaloids, aplaminone, neoaplaminone, and neoaplaminone sulfate from the marine mollusc *Aplysia kurodai*. Tetrahedron Lett. 34:4911–4914.

Kinnel, R.B., Dieter, R.K., Meinwald, J., Van Engen, D., Clardy, J., Eisner, T., Stallard, M.O., and Fenical, W. 1979. Brasilenyne and *cis*-dihydrorhodophytin: antifeedant medium-ring haloethers from a sea hare (*Aplysia brasiliana*). Proc. Natl. Acad. Sci. USA 76:3576–3579.

Kinnel, R., Duggan, A.J., Eisner, T., Meinwald, J., and Miura, I. 1977. Panacene: an aromatic bromoallene from a sea hare (*Aplysia brasiliana*). Tetrahedron Lett. 44:3913–3916.

Ksebati, M.B., and Schmitz, F.J. 1985. Tridachiapyrones: propionate-derived metabolites from the sacoglossan mollusc *Tridachia crispata*. J. Org. Chem. 50:5637–5642.

Kubo, I., Lee, Y.-W., Pettei, M., Pilkiewicz, F., and Nakanishi, K. 1976. Potent armyworm antifeedants from the East African Warburgia plants. J. Chem. Soc. Chem. Commun. 1013–1014.

Kusumi, T., Uchida, H., Inouye, Y., Ishitsuka, M., Yamamoto, H., and Kakisawa, H. 1987. Novel cytotoxic monoterpenes having a halogenated tetrahydropyran from *Aplysia kurodai*. J. Org. Chem. 52:4597–4600.

Lewin, R.A. 1970. Toxin secretion and tail autotomy by irritated *Oxynoe panamensis* (Opisthobranchiata; Sacoglossa). Pac. Sci. 24:356–358.

Lindquist, N.L. 1989. Secondary metabolite production and chemical adaptations in the class Ascidiacea. Ph.D. dissertation, University of California, San Diego.

Lindquist, N., Fenical, W., Van Duyne, G.D., and Clardy, J. 1988. New alkaloids of the lamellarin class from the marine ascidian *Didemnum chartaceum* (Sluiter, 1909). J. Org. Chem. 53:4570–4574.

McDonald, F.J., Campbell, D.C., Vanderah, D.J., Schmitz, F.J., Washecheck, D.M., Burks, J.E., and van der Helm, D. 1975. Marine natural products. Dactylyne, an acetylenic dibromochloro ether from the sea hare *Aplysia dactylomela*. J. Org. Chem. 40:665–666.

Manker, D.C. 1988. Occurrence, origin and function of secondary metabolites in marine pulmonate molluscs. Ph.D. dissertation, University of California, San Diego.

Manker, D.C., and Faulkner, D.J. 1989. Vallartanones A and B, polypropionate metabolites of *Siphonaria maura* from Mexico. J. Org. Chem. 54:5374–5377.

Manker, D.C., Garson, M.J., and Faulkner, D.J. 1988. *De novo* biosynthesis of polypropionate metabolites in the marine pulmonate *Siphonaria denticulata*. J. Chem. Soc. Chem. Commun. 1061–1062.

Matsuda, H., Tomiie, Y., Yamamura, S., and Hirata, Y. 1967. The structure of aplysin-20. Chem. Commun., pp. 898–899.

Matsunaga, S., Fusetani, N., Hashimoto, K., Koseki, K., and Noma, M. 1986. Kabiramide C, a novel antifungal macrolide from nudibranch eggmasses. J. Am. Chem. Soc. 108:847–849.

Mebs, D. 1985. Chemical defense of a dorid nudibranch, *Glossodoris quadricolor*, from the Red Sea. J. Chem. Ecol. 11:713–716.

Midland, S.L., Wing, R.M., and Sims, J.J. 1983. New crenulides from the sea hare *Aplysia vaccaria*. J. Org. Chem. 48:1906–1909.

Minale, L., and Riccio, R. 1976. Constituents of the digestive gland of the molluscs of the genus *Aplysia*. I. Novel diterpenes from *Aplysia depilans*. Tetrahedron Lett., pp. 2711–2714.

Miyamoto, T., Higuchi, R., Komori, T., Fujioka, T., and Mihashi, K. 1986. Isolation and structures of aplykurodins A and B, two new isoprenoids from the marine mollusk *Aplysia kurodai*. Tetrahedron Lett. 27:1153–1156.

Miyamoto, T., Higuchi, R., Marubyashi, N., and Komori, T. 1988. Studies on the constituents of marine opisthobranchia. IV. Two new polyhalogenated monoterpenes from the sea hare *Aplysia kurodai*. Liebigs Ann. Chem., pp. 1191–1193.

Moore, R.E., Blackman, A.J., Cheuk, C.E., Mynderse, J.S., Matsumoto, G.K., Clardy, J.,

Woodward, R.W., and Craig, J.C. 1984. Absolute stereochemistries of the aplysiatoxins and oscillatoxin A. J. Org. Chem. 49:2484–2489.

Neèman, L.B., Fischelson, L., and Kashman, Y. 1975. Isolation of a new toxin from the sponge *Latrunculia magnifica* in the Gulf of Aquaba (Red Sea). Mar. Biol. 30:293–296.

Ojika, M., Yoshida, Y., Nakayama, Y., and Yamada, K. 1990a. Aplydilactone, a novel fatty acid metabolite from the marine mollusc *Aplysia kurodai*. Tetrahedron Lett. 34:4907–4910.

Ojika, M., Yoshida, Y., Okumura, M., Ieda, S., and Yamada, K. 1990b. Aplysiadiol, a new brominated diterpene from the marine mollusc *Aplysia kurodai*. J. Nat. Prod. 53:1619–1622.

Okuda, R.K., and Scheuer, P.J. 1985. Latrunculin-A, ichthyotoxic constituent of the nudibranch *Chromodoris elisabethina*. Experientia 41:1355–1356.

Okuda, R.K., Scheuer, P.J., Hochlowski, J.E., Walker, R.P., and Faulkner, D.J. 1983. Sesquiterpenoid constituents of eight porostome nudibranchs. J. Org. Chem. 48:1866–1870.

Paine, R.T. 1963. Food recognition and predation on opisthobranchs by *Navanax inermis*. Veliger 6:1.

Paul, V.J., Lindquist, N., and Fenical, W. 1990. Chemical defenses of the tropical ascidian *Atapozoa* sp. and its nudibranch predators *Nembrotha* spp. Mar. Ecol. Prog. Ser. 59:109–118.

Paul, V.J., and Van Alstyne, K.L. 1988. Use of ingested algal diterpenoids by *Elysia halimedae* Macnae (Opisthobranchia: Ascoglossa) as antipredator defenses. J. Exp. Mar. Biol. Ecol. 119:15–29.

Pawlik, J.R., Albizati, K.F., and Faulkner, D.J. 1986. Evidence of a defensive role for limatulone, a novel triterpene from the intertidal limpet *Collisella limatula*. Mar. Ecol. Prog. Ser. 30:251–260.

Pawlik, J.R., Kernan, M.R., Molinski, T.F., Harper, M.K., and Faulkner, D.J. 1988. Defensive chemicals of the Spanish dancer nudibranch *Hexabranchus sanguineus* and its egg ribbons: macrolides derived from a sponge diet. J. Exp. Mar. Biol. Ecol. 119:99–109.

Pennings, S.C. 1990. Multiple factors promoting narrow host range in the sea hare, *Aplysia californica*. Oecologia 82:192–200.

Pettit, G.R., Herald, C.L., Allen, M.S., Von Dreele, R.B., Vanell, L.D., Kao, J.P.Y., and Blake, W. 1977. The isolation and structure of aplysiastatin. J. Am. Chem. Soc. 99:262–263.

Pettit, G.R., Herald, C.L., Einck, J.J., Vanell, L.D., Brown, P., and Gust, D. 1978. Isolation and structure of angasiol. J. Org. Chem. 43:4685–4686.

Pettit, G.R., Kamano, Y., Brown, P., Gust, D., Inoue, M., and Herald, C. L. 1982. Structure of the cyclic peptide dolastatin 3 from *Dolabella auricularia*. J. Am. Chem. Soc. 104:905–907.

Pettit, G.R., Kamano, Y., Dufresne, C., Cerny, R.L., Herald, C.L., and Schmidt, J.M. 1989c. Isolation and structure of the cytostatic linear depsipeptide dolastatin 15. J. Org. Chem. 54:6005–6006.

Pettit, G.R., Kamano, Y., Dufresne, C., Herald, C.L., Bontems, R.J., Schmidt, J.M., Boettner, F.E., and Nieman, R.A. 1989a. Isolation and structure of the cell growth inhibitory depsipeptides dolastatins 11 and 12. Heterocycles 28:553–557.

Pettit, G.R., Kamano, Y., Herald C.L., Dufresne, C., Bates, R.B., Schmidt, J.M., Cerny, R.L., and Kizu, H. 1990. Antineoplastic agents. 190. Isolation and structure of the cyclodepsipeptide dolastatin 14. J. Org. Chem. 55:2989–2990.

Pettit, G.R., Kamano, Y., Herald, C.L., Dufresne, C., Cerny, R.L., Herald, D.L., Schmidt, J.M., and Kizu, H. 1989b. Isolation and structure of the cytostatic depsipeptide dolastatin 13 from the sea hare *Dolabella auricularia*. J. Am. Chem. Soc. 111:5015–5017.

Pettit, G.R., Kamano, Y., Herald, C.L., Tuinman, A.A., Boettner, F.E., Kizu, H., Schmidt, J.M., Baczynskyj, L., Tomer, K.B., and Bontems, R.J. 1987a. The isolation and structure of a remarkable marine animal antineoplastic constituent: dolastatin 10. J. Am. Chem. Soc. 109:6883–6885.

Pettit, G.R., Kamano, Y., Holzapfel, C.W., van Zyl, W.J., Tuinman, A.A., Herald, C.L., Baczynskyj, L., and Schmidt, J.M. 1987b. The structure and synthesis of dolastatin 3. J. Am. Chem. Soc. 109:7581–7582.

Pettit, G.R., Ode, R.H., Herald, C.L., Von Dreele, R.B., and Michel, C. 1976. The isolation and structure of dolatriol. J. Am. Chem. Soc. 98:4677–4678.

Poiner, A., Paul, V.J., and Scheuer, P. J. 1989. Kumepaloxane, a rearranged trisnor sesquiterpene from the bubble shell *Haminoea cymbalum*. Tetrahedron 45:617–622.

Quiñoa, E., Castedo, L., and Riguera, R. 1989. The halogenated monoterpenes of *Aplysia punctata*. A comparative study. Comp. Biochem. Physiol. B 92:99–101.

Rice, S.H. 1985. An anti-predator chemical defense of the marine pulmonate gastropod *Trimusculus reticulatus* (Sowerby). J. Exp. Mar. Biol. Ecol. 93:83–89.

Roesener, J.A., and Scheuer, P.J. 1986. Ulapualide A and B, extraordinary antitumor macrolides from nudibranch eggmasses. J. Am. Chem. Soc. 108:846–847.

Rose, A.F., Scheuer, P.J., Springer, J. P., and Clardy, J., 1978. Stylocheilamide, an unusual constituent of the sea hare *Stylocheilus longicauda*. J. Am. Chem. Soc. 100:7665–7670.

Roussis, V., Pawlik, J.R., Hay, M.E., and Fenical, W. 1990. Secondary metabolites of the chemically rich ascoglossan *Cyerce nigricans*. Experientia 46:327–329.

Rüdiger, W. 1967a. Über die Abwehrfarbstoffe von *Aplysia*—Artin I, Aplysioviolin, ein neuartiger Gallenfarbstoff. Hoppe-Seyler's Z. Physiol. Chem. 348:129–138.

Rüdiger, W. 1967b. Über die Abwehrfarbstoffe von *Aplysia*—Artin II, die Struktur von Aplysioviolin. Hoppe-Seyler's Z. Physiol. Chem. 348:1554.

Sakai, R., Higa, T., Jefford, C. W., and Bernardinelli, G. 1986. The absolute configurations and biogenesis of some new halogenated chamigrenes from the sea hare *Aplysia dactylomela*. Helv. Chim. Acta 69:91–105.

Schmitz, F.J., Hollenbeak, K.H., Carter, D.C., Hossain, M.B., and van der Helm, D. 1979. Marine natural products: 14-bromoobtus-1-ene-3,11-diol, a new diterpenoid from the sea hare *Aplysia dactylomela*. J. Org. Chem. 44:2445–2447.

Schmitz, F.J., Hollenbeak, K.H., and Vanderah, D.J. 1978b. Marine natural products: dactylol, a new sesquiterpene alcohol from a sea hare. Tetrahedron 34:2719–2722.

Schmitz, F.J., and McDonald, F.J. 1974. Marine natural products: dactyloxene-B, a sesquiterpene ether from the sea hare, *Aplysia dactylomela*. Tetrahedron Lett. 2541–2544.

Schmitz, F.J., McDonald, F.J., and Vanderah, D.J. 1978a. Marine natural products: sesquiterpene alcohols and ethers from the sea hare *Aplysia dactylomela*. J. Org. Chem. 43:4220–4225.

Schmitz, F.J., Michaud, D.P., and Hollenbeak, K.H. 1980. Marine natural products: dihydroxydeodactol monoacetate, a halogenated sesquiterpene ether from the sea hare *Aplysia dactylomela*. J. Org. Chem. 45:1525–1528.

Schmitz, F.J., Michaud, D.P., and Schmidt, P.G. 1982. Marine natural products: parguerol, deoxyparguerol, and isoparguerol. New brominated diterpenes with modified pimarane skeletons from the sea hare *Aplysia dactylomela*. J. Am. Chem. Soc. 104:6415–6423.

Schulte, G.R., Chung, M.C.H., and Scheuer, P.J. 1981. Two bicyclic C_{15} enynes from the sea hare *Aplysia oculifera*. J. Org. Chem. 46:3870–3873.

Schulte, G., Scheuer, P.J., and McConnell, O.J. 1980. Two furanosesquiterpene marine metabolites with antifeedant properties. Helv. Chim. Acta 63:2159–2167.

Sleeper, H.L., and Fenical, W. 1977. Navenones A–C: trail-breaking alarm pheromones from the marine opisthobranch *Navanax inermis*. J. Am. Chem. Soc. 99:2367–2368.

Sleeper, H.L., Paul, V.J., and Fenical, W. 1980. Alarm pheromones from the marine opisthobranch *Navanax inermis*. J. Chem. Ecol. 6:57–70.

Stallard, M.O. 1974. Chemical constituents of the sea hare *Aplysia californica*. Ph.D. dissertation, University of California, San Diego.

Stallard, M.O., and Faulkner, D.J. 1974a. Chemical constituents of the digestive gland of the sea hare *Aplysia californica*. I. Importance of diet. Comp. Biochem. Physiol. B 49:25–35.

Stallard, M.O., and Faulkner, D.J. 1974b. Chemical constituents of the digestive gland of the sea hare *Aplysia californica*. II. Chemical transformations. Comp. Biochem. Physiol. B 49:37–41.

Stallard, M.O., Fenical, W., and Kittredge, J.S. 1978. The brasilenols, rearranged sesquiterpene alcohols isolated from the marine opisthobranch *Aplysia brasiliana*. Tetrahedron 34:2077–2081.

Thompson, J.E., Walker, R.P., and Faulkner, D.J. 1985. Screening and bioassays for biologically-active substances from 40 marine sponge species from San Diego, California. Mar. Biol. 88:11–21.

Thompson, J.E., Walker, R.P., Wratten S.J., and Faulkner, D.J. 1982. A chemical defense mechanism for the nudibranch *Cadlina luteomarginata*. Tetrahedron 38:1865–1873.

Thompson, T.E. 1960. Defensive adaptations in opisthobranchs. J. Mar. Biol. Assoc. U.K. 39:123–134.

Thompson, T.E. 1969. Acid secretion in Pacific Ocean gastropods. Aust. J. Zool. 17:755–764.

Thompson, T.E. 1983. Detection of epithelial acid secretions in marine molluscs: review of techniques, and new analytical methods. Comp. Biochem Physiol. A 74:615–621.

Vanderah, D.J., and Schmitz, F. J. 1976. Marine natural products: isodactylyne, a halogenated actylenic ether from the sea hare *Aplysia dactylomela*. J. Org. Chem. 41:3480–3481.

Walker, R.P. 1981. The chemical ecology of some sponges and nudibranchs from San Diego. Ph.D. dissertation, University of California, San Diego.

Walsby, J.R., Morton, J.E., and Croxall, J.P. 1973. The feeding mechanism and ecology of the New Zealand pulmonate limpet *Gadinalea nirea*. J. Zool. Lond. 171:257–283.

Watson, M. 1973. Midgut gland toxins of Hawaiian sea hares. I. Isolation and preliminary toxicological observations. Toxicon 11:259–267.

Winkler, L.R. 1961. Preliminary tests of the toxin extracted from California sea hares of the genus *Aplysia*. Pac. Sci. 15:211–214.

Yamamura, S., and Terada, Y. 1977. Isoaplysin-20, a natural bromine-containing di-terpene, from *Aplysia kurodai*. Tetrahedron Lett., pp. 2171–2172.

Yonge, C.M. 1957. Observations in life on the pulmonate limpet *Trimusculus* (*Ga-dinia*) *reticulatus* (Sowerby). Proc. Malacol. Soc. Lond. 33:31–38.

Young, C.M., Greenwood, P.G., and Powell, C.J. 1986. The ecological role of defensive secretions in the intertidal pulmonate *Onchidella borealis*. Biol. Bull. 171:391–404.

Chapter 5
Chemical Defenses of Benthic Marine Invertebrates

VALERIE J. PAUL

A variety of sessile marine invertebrates inhabit benthic substrates in temperate and tropical oceans. Several groups of these invertebrates, including sponges, coelenterates, bryozoans, and ascidians, are found only in aquatic, and primarily marine, habitats. Some of the coelenterates, particularly stony corals, and some bryozoans secrete massive calcium carbonate ̓skeletons; some coelenterates and sponges produce spicules or sclerites. Many of these organisms, however, are soft-bodied and relatively vulnerable to predators, and chemical defenses may be important to them. Thousands of secondary metabolites have been isolated from marine invertebrates such as sponges, soft corals, gorgonian corals, zoanthids, bryozoans, and ascidians. The predominance of these metabolites in sessile invertebrates that inhabit areas of intense predation pressure—and have few or no apparent morphological defenses—suggests a defensive function for these chemicals.

Evidence supporting the role of secondary metabolites as chemical defenses in marine invertebrates is limited but accumulating rapidly. Results of studies investigating the functions of invertebrate secondary metabolites as defenses against predators, competitors, and fouling organisms have been recently reviewed (Sammarco and Coll 1988; Davis et al. 1989) and are discussed in this chapter. I focus on chemical defenses in marine benthic invertebrates because these organisms have been studied the most by natural products chemists and ecologists. Recent studies suggest that pelagic invertebrates also possess chemical defenses (Shanks and Graham 1988; McClintock and Janssen 1990); these interactions clearly warrant further research.

Ecology and Chemical Defense against Predation

Predation intensity in marine benthic communities is thought to be greatest in the tropics, particularly in the Indo-Pacific region (Bakus 1964, 1969; Vermeij 1978, 1987; Bertness et al. 1981; Gaines and Lubchenco 1982; Steneck 1986). Predators and herbivores play a major role in structuring both temperate and tropical benthic communities, and many tropical organisms seem to be particularly well defended chemically and morphologically (Bakus and Green 1974; Green 1977; Vermeij 1978; Steneck 1986, 1988; Hay and Fenical 1988). The tropical organisms most extensively investigated by natural products chemists include seaweeds, octocorals (soft corals and gorgonian corals), sponges, ascidians, and echinoderms (sea stars and sea cucumbers). This chapter emphasizes predation on sessile invertebrates (octocorals, sponges, and ascidians), most of which are preyed on by relatively few specialized predators.

Soft corals (Alcyonacea) and gorgonian corals (Gorgonacea) are most abundant and diverse in tropical habitats. Soft corals live primarily in the tropical Indo-Pacific region; they are rare in the eastern Pacific and the Caribbean. Gorgonian corals are found in temperate and tropical seas but are most abundant and diverse in shallow Caribbean waters (Bayer 1961). Neither group has an external calcium carbonate skeleton like that of the stony corals, but both groups have calcitic sclerites (calcium carbonate spicules) embedded in their coenenchyme that are thought to serve as structural support for the colony (Koehl 1982; Harvell and Fenical 1989). A defensive role for sclerites in soft corals was suggested by Sammarco et al. (1987), and sclerites from *Pseudopterogorgia acerosa* (gorgonian) and *Sinularia* spp. (soft corals) have been demonstrated to deter predators in field assays (Harvell et al. 1988; Van Alstyne et al. 1992).

The major group of generalist predators on soft corals and gorgonians are the butterflyfish (Chaetodontidae). Although most butterflyfish feed primarily on hard corals, some species consume large amounts of soft corals or gorgonians during certain times of the year or at certain locations (Randall 1967, 1974; Hobson 1974; Anderson et al. 1981; Birkeland and Neudecker 1981; Harmelin-Vivien and Bouchon-Navaro 1981, 1983; Tursch and Tursch 1982; Lasker 1985; Harvell et al. 1988; Wylie and Paul 1989). A few other fishes also consume octocorals.

Pawlik et al. (1987) and Harvell et al. (1988) reported that the common bluehead wrasse (*Thalassoma bifasciatum*) will pick at gorgonian corals in the Caribbean. The scrawled filefish (*Alutera scripta*) has also been reported to consume gorgonian polyps in Belize (Harvell et al. 1988).

The polychaete fireworm (*Hermodice carunculata*) and several groups of molluscs eat primarily octocorals. These animals remove tissue from localized areas of the colonies and can easily kill young colonies. *H. carunculata* shows distinct preferences for some species of gorgonians in the Caribbean and seems to prefer *Briareum asbestinum* (Vreeland and Lasker 1989). The flamingo-tongue cowry (*Cyphoma gibbosum*) eats most species of gorgonians in the Caribbean. Preferences for different species of gorgonians appear to vary among individual cowries and among locations in the Caribbean. Some researchers have reported distinct feeding preferences for these molluscs, while others have found them randomly distributed on gorgonians based on relative abundances of different gorgonian species (Kinzie 1970; Birkeland and Gregory 1975; Hazlett and Bach 1982; Gerhart 1986; Harvell and Suchanek 1987; Lasker et al. 1988; Lasker and Coffroth 1988). Another mollusc, the egg cowry (*Ovula ovum*), feeds on soft corals in the Indo-Pacific. This mollusc prefers species of *Sarcophyton* and can transform sarcophytoxide (1), a major diterpenoid metabolite found in its preferred prey, to a less toxic metabolite, 7,8-deoxysarcophytoxide (2) (Coll et al. 1983).

1   2

Opisthobranch molluscs such as nudibranchs also feed on octocorals in both tropical and temperate habitats. Nudibranchs are among the most specialized predators on the octocorals, and most species feed on just one or a few related octocoral species (Willan and Coleman 1984; Gosliner 1987).

Fishes, including angelfishes and filefishes, hawksbill turtles, and molluscs (chitons, opisthobranchs) are among the major sponge predators in the tropics. Randall and Hartman (1968) reported that sponges constituted more than 95% of the food of angelfishes of the genus

Holacanthus, more than 70% of the food of angelfishes of the genus *Pomacanthus*, and more than 85% of the food of the filefish *Cantherhines macrocerus*, based on stomach content analyses. However, feeding on sponges is relatively rare in other fishes: of 212 species of reef fishes in the West Indies, sponges were found in the stomachs of 21 species. Sponges made up more than 6% of the stomach contents of only 11 species of fish; all were angelfish, spadefish, filefish, or trunkfish. Some of these species appear to specialize on sponges and consume only small amounts of other prey items. The hawksbill turtle (*Eretmochelys imbricata*), an endangered species found in the tropics, eats sponges almost exclusively in the Caribbean and possibly in other areas (Meylan 1988). Opisthobranch molluscs such as nudibranchs are common specialist predators on sponges in temperate and tropical habitats (Bakus and Abbott 1980; Willan and Coleman 1984; Gosliner 1987). Sea stars may be important sponge predators in Antarctic benthic communities (McClintock 1987; Dayton 1989).

Benthic ascidians seem to be relatively free of generalist predators (Millar 1971; Goodbody and Gibson 1974; Stoecker 1980c). In the tropics, some angelfishes, spadefishes, filefishes, and pufferfishes are known to consume relatively large amounts of ascidians (Randall and Hartman 1968; Myers 1983). Molluscs such as *Trochus niloticus* and *Turbo* spp. have been observed feeding on ascidians on the Great Barrier Reef (Parry 1984). Specialized predators on ascidians include polyclad flatworms and molluscs such as lamellarians, cypraeids (cowries), and nudibranchs (Millar 1971; Abbott and Newberry 1980; Parry 1984). Similarly, many tropical bryozoans have few generalist predators, although some nudibranchs, especially members of the family Polyceridae, specialize on them (Willan and Coleman 1984; Gosliner 1987).

Because of the few known generalist predators on these groups of invertebrates, it has been assumed that secondary metabolites play a major role in their defense. Many invertebrate secondary metabolites are cytotoxic, antimicrobial, or have other biological activities in laboratory assays (Ireland et al. 1988). Many soft corals are toxic freshwater mosquito fish (*Gambusia affinis*) when fish are exposed to seawater containing homogenates of the corals (Coll et al. 1982b; La Barre et al. 1986b). These same soft coral metabolites may or may not be toxic when ingested by coral reef fishes. There are few realistic assessments of the deterrent functions of invertebrate secondary metabolites against potential predators.

Several studies have demonstrated a feeding deterrent role for octocoral extracts and secondary metabolites. Gerhart (1984) showed that prostaglandins in the gorgonian *Plexaura homomalla* induced regurgitation in the yellow-head wrasse (*Halichoeres garnoti*) in field assays. The compounds used in his assays were the hydroxy acids of 15(S) and 15(R) prostaglandin A_2 (3); however, Pawlik and Fenical (1989) later pointed out that acetoxy methyl esters (4), not the hydroxy acids, are the actual natural products stored in *P. homomalla* tissues. The acetoxy methyl esters were not effective feeding deterrents in field assays (Pawlik and Fenical 1989), although the hydroxy acid and partially esterified prostaglandins, such as the hydroxy methyl ester (5) and acetoxy acid (6),

3 HYDROXY ACID: $R_1 = H$, $R_2 = H$
4 ACETOXY METHYL ESTER: $R_1 = CH_3$, $R_2 = COCH_3$
5 HYDROXY METHYL ESTER: $R_1 = CH_3$, $R_2 = H$
6 ACETOXY ACID: $R_1 = H$, $R_2 = COCH_3$

were all deterrents. Thus, results concerning the deterrent effects of prostaglandins in *P. homomalla* are contradictory; the importance of the prostaglandins as feeding deterrents appears to depend on the rate at which fully esterified forms of the prostaglandins are hydrolyzed after damage to gorgonian tissues. Based on previous evidence, this rate appears to be slow (up to 24 hours; Schneider et al. 1977).

Pawlik et al. (1987) tested organic extracts of Caribbean gorgonians against the wrasse *Thalassoma bifasciatum* in aquarium assays and found that 51% of the extracts were highly unpalatable. Crude lipid extracts of two Caribbean gorgonians (*Pseudopterogorgia acerosa* and *P. rigida*), two isolated metabolites from *P. rigida*, curcuhydroquinone (7) and curcuquinone (8),

7   8

and calcitic sclerites from *P. acerosa* effectively reduced predation on palatable food strips in field assays (Harvell et al. 1988). Gerhart et al.

(1988) showed that spicules but not extracts of the temperate gorgonian *Leptogorgia virgulata* deterred pinfish (*Lagodon rhomboides*) when incorporated into alginate pellets in aquarium assays. A natural combination of spicules and extracts was particularly effective in reducing acceptance of pellets by pinfish. On Guam, extracts of three species of the soft coral *Sinularia* and a major cembranoid diterpene (9)

9

from *S. maxima* deterred generalist fishes in field assays (Wylie and Paul 1989). These same extracts and the isolated metabolite were ineffective except at very high concentrations (20% dry weight) against the butterflyfish *Chaetodon unimaculatus*, a major predator on *Sinularia* species on Guam (Wylie and Paul 1989).

Van Alstyne and Paul (1992) demonstrated that the common Caribbean sea fan (*Gorgonia ventalina*) uses both calcified sclerites and secondary metabolites as defenses against predators. Crude organic extracts and sclerites at natural concentrations significantly deterred feeding by natural assemblages of reef fishes in field assays. The cowrie (*Cyphoma gibbosum*), a gorgonian specialist, also avoided diets containing extracts and sclerites of *G. ventalina* at natural concentrations in shipboard assays. In addition, fractions containing the nonpolar sesquiterpenes from *G. ventalina* deterred feeding by fishes in field assays.

Defensive properties have also been demonstrated for secondary metabolites produced by the Caribbean gorgonian *Erythropodium caribaeorum* (Fenical and Pawlik 1991). Crude extracts deterred feeding by natural populations of reef fishes in field assays. The deterrent activity was found in a fraction of the extract that contained chlorinated diterpenes known as erythrolides; another fraction containing sesquiterpene hydrocarbons was not deterrent. Several of the isolated erythrolides, when tested individually at natural concentrations, deterred feeding by reef fishes.

Several studies have shown that concentrations of extracts and secondary metabolites are highest in the tip portions of gorgonians and soft corals (Harvell and Fenical 1989; Wylie and Paul 1989). These portions of the colonies are often the major polyp-bearing regions of the colonies, and they have lower concentrations of sclerites and are more exposed to predators than the basal portions. These observations suggest that chemical defenses may be allocated to the most vulnerable parts of the colonies. The pattern of higher concentrations of extracts and secondary metabolites where concentrations of sclerites are low does not occur in all gorgonians. Concentrations of sclerites and extracts vary little within the main body of *G. ventalina* colonies, although they are both significantly lower in the basal stems (Van Alstyne and Paul 1991).

The deterrent roles of sponge secondary metabolites are even less well studied than those of octocorals. Many studies indicate that extracts of sponges from tropical, temperate, and Antarctic waters have antimicrobial or ichthyotoxic activities (Bakus and Green 1964; Green 1977; Bakus 1981; Amade et al. 1982; Mebs et al. 1985; McCaffrey and Endean 1985; McClintock 1987); however, only a few studies have demonstrated a feeding deterrent role for these compounds (Thompson et al. 1985; Pawlik et al. 1988). My colleagues and I have tested extracts of about a dozen tropical sponges and several of their major metabolites as feeding deterrents against generalist fishes in field assays on Guam (Rogers and Paul 1991; Duffy and Paul 1992; Paul, pers. observ.). Isolated sponge metabolites tested included avarol (10), heteronemin (11), scalardial (12), latrunculin-A (13), manoalide (14), secomanoalide (15), and a brominated diphenyl ether (16) from *Dysidea* sp.

10   11

All metabolites except latrunculin-A deterred generalist fishes at or below natural concentrations. Latrunculin-A did deter the pufferfish *Canthigaster solandri* in aquarium assays. Scalardial deterred natural

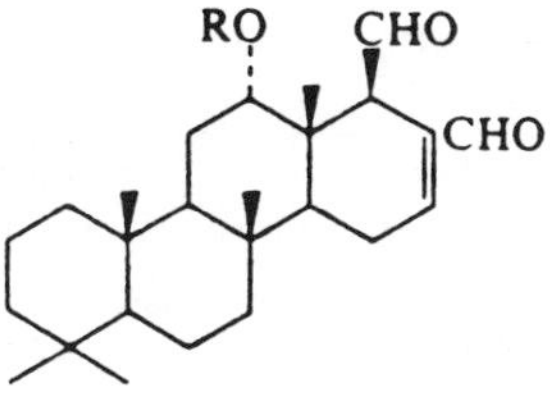

12 R = Ac

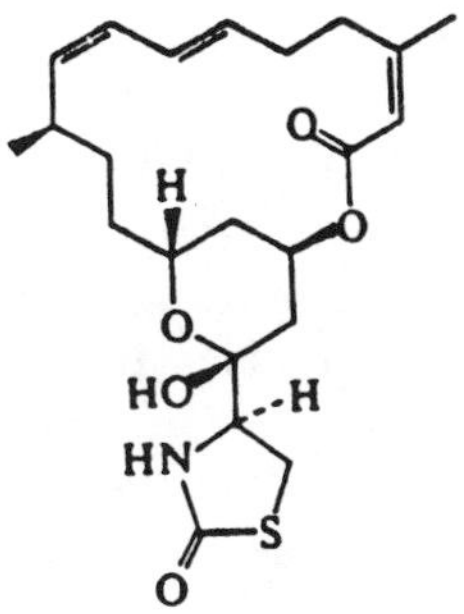

13

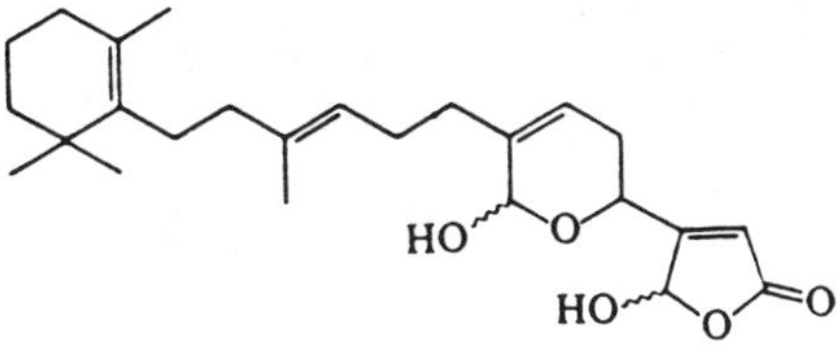

14

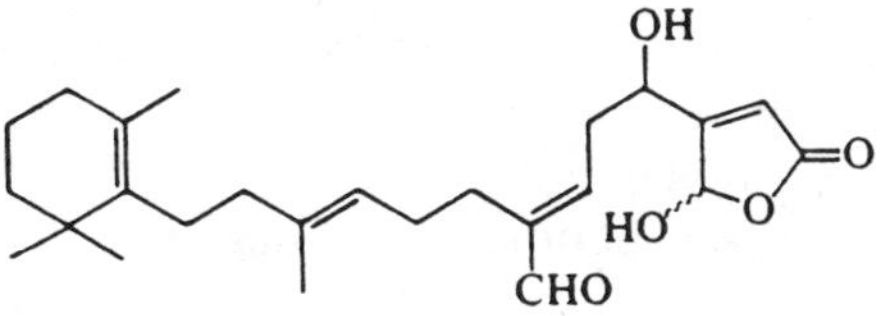

15

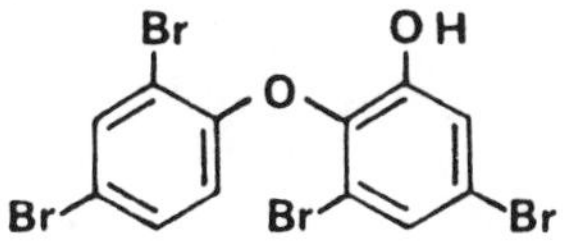

16

populations of fishes on some Guam reefs but not others (Rogers and Paul 1991). In some cases, natural mixtures of metabolites were more effective deterrents than single pure metabolites when tested at the same concentrations. Most sponge extracts, which contained mixtures of secondary metabolites, were significant feeding deterrents even though some isolated metabolites from the same sponges were not deterrent. Thus, our results with sponge secondary metabolites were similar to those with secondary metabolites from other organisms (e.g., seaweeds and coelenterates): some isolated metabolites strongly deterred generalist fishes while others did not.

Benthic ascidians have been hypothesized to use high vanadium concentrations and low pH as well as secondary metabolites as defenses against predation and fouling (Stoecker 1978; 1980a, 1980b, 1980c). Parry (1984), however, proposed that neither pH nor vanadium prevents predation on ascidians. The vanadium concentrations of most of the species he examined were low, though, and acid is rapidly neutralized when cell damage occurs. Some evidence for chemical defenses in ascidians has been observed. The larvae of the ascidians *Trididemnum solidum, Didemnum molle,* and *Ecteinascidia turbinata* were rejected by fish predators (van Duyl et al. 1981; Olson 1983; Young and Bingham 1987). Larvae of *E. turbinata* seemed to be defended by low-molecular-weight organic metabolites, although no deterrent secondary metabolites were actually isolated (Young and Bingham 1987). Young and Bingham (1987) also suggested that chemical defenses are present in other invertebrate larvae, including the ascidian *Eudistoma olivaceum,* the sea cucumber *Psolus chitonoides,* and the octocoral *Briareum asbestinum.*

McClintock et al. (1991) showed that the tunic of the Antarctic ascidian *Cnemidocarpa verrucosa* was distasteful to potential fish predators and likely contained chemical defenses. No specific secondary metabolites were isolated, however, to explain the tunic's deterrent properties.

Paul et al. (1990) demonstrated that secondary metabolites produced by the tropical western Pacific ascidian *Sigillina signifera* (= *Atapozoa* sp.) were significant feeding deterrents against predatory fishes in field assays on Guam. The ascidian and its nudibranch predators (*Nembrotha* spp.) contained high concentrations of a series of bipyrrole metabolites known as the tambjamines (Carté and Faulkner 1983). The crude extract, mixtures of tambjamines, tambjamine C (17),

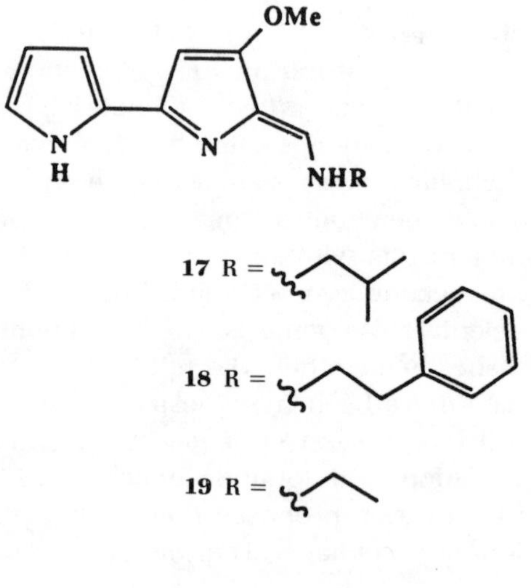

17 R =

18 R =

19 R =

and tambjamine F (18) were all significant deterrents at or below natural concentrations, while tambjamine E (19) was not deterrent. Both the ascidian and its nudibranch predators appear to use the tambjamines as chemical defenses against predators. Tambjamines C and E were also found in larvae of *Sigillina* (Lindquist 1989; Lindquist and Fenical 1991), which suggests that the larvae may be chemically defended as well.

Tambjamines are also found in the bryozoans *Sessibugula translucens* and *Bugula dentata* (Carté and Faulkner 1983, 1986; Lindquist and Fenical 1991). These bryozoans are also eaten by nembrothid nudibranchs of the genus *Tambja*. It is interesting that tambjamines are found in both bryozoans and ascidians and that nembrothid nudibranchs appear to seek out these metabolites, whatever the source. Carté and Faulkner (1986) demonstrated a feeding deterrent role for tambjamines isolated from *Sessibugula translucens* and *Tambja* spp. from the Gulf of California. Mixtures of halogenated and non-halogenated tambjamines deterred the California spotted kelpfish (*Gibbonsia elegans*) in aquarium assays.

Lindquist (1989) recently tested a variety of ascidian secondary metabolites in an agar-based diet in field assays using generalist fishes. In his tests conducted at Looe Key, Florida, with metabolites from Caribbean ascidians, didemnenones A (20) and B (21)

20 $R_1 = H$, $R_2 = OH$, $R_3 = OH$
21 $R_1 = OH$, $R_2 = H$, $R_3 = OH$

from *Trididemnum* cf. *cyanophorum*, and spicules from *Trididemnum solidum* were not deterrent at natural concentrations. Didemnin B (22) and nordidemnin B (23)

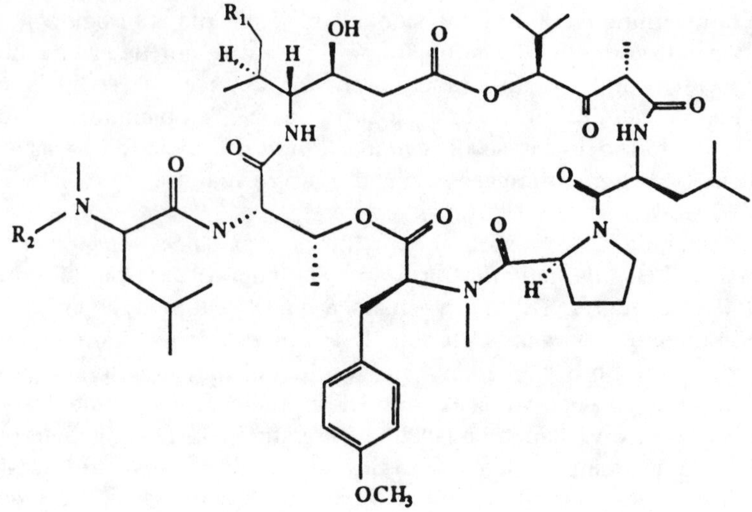

22 R_1 = Me

R_2 = $CH_3\overset{L}{C}HOHCO$ – $N\overset{L}{}CO$

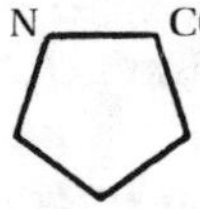

23 R_1 = H

R_2 = $CH_3\overset{L}{C}HOHCO$ – $N\overset{L}{}CO$

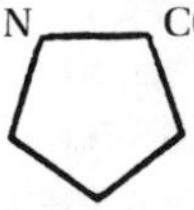

from *Trididemnum solidum* were highly deterrent even at one-tenth their natural concentration. Metabolites from two Indo-Pacific ascidians, patellamide C (24)

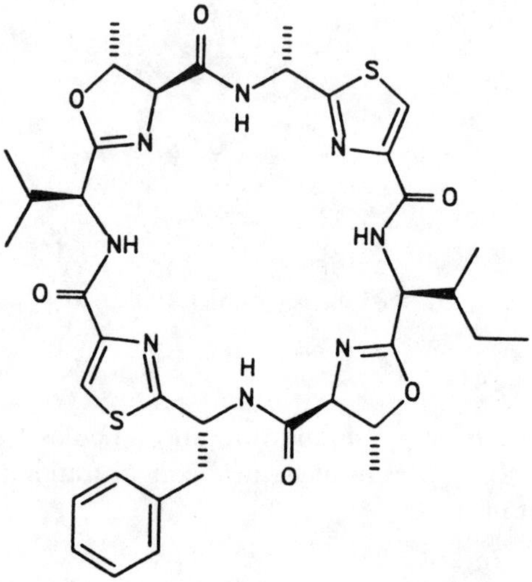

24

from *Lissoclinum patella* and tambjamine E (19) from *Atapozoa* sp., were also tested in field assays in the Philippines; tambjamine E was a significant deterrent at natural concentrations, while patellamide C was not. A mixture of metabolites from *Polyandrocarpa* sp., polyandrocarpidines A–D (25–28),

25 n = 3
27 n = 2

26 n = 3
28 n = 2

from the Gulf of California was tested against five invertebrate predators in aquarium assays. The mixture of metabolites was deterrent to four of the five invertebrates species. Lindquist's results support other observations that even closely related metabolites affect predators differently (Paul et al. 1988, 1990; Hay et al. 1988). Some compounds are broadly deterrent to a variety of predators, while other compounds show little or no deterrent effect.

Clearly, only a few secondary metabolites from marine invertebrates have been examined for deterrent functions in ecologically realistic assays with potential natural predators. Hundreds of secondary metabolites have been isolated from sponges, octocorals, ascidians, and, to a lesser extent, bryozoans and zoanthids. Considering the variation in effects of the few compounds that have been tested against the diversity of predators present in the field, many questions remain regarding the defensive role of these compounds with regard to generalist and specialist predators. How do secondary metabolite concentrations and types vary within and among individual organisms? Can predators or environmental factors influence the production of secondary metabolites? For example, Thompson et al. (1987) demonstrated that different field conditions of illumination and depth affected the diterpene composition of the sponge *Rhopaloeides odorabile*. How do secondary metabolites affect the physiology of generalist and specialist consumers? Additionally, secondary metabolites in marine invertebrates may mediate other behavioral interactions such as escape and avoidance by prey organisms in response to chemicals released by their predators. The chemical bases for these interactions have rarely been investigated (Fishlyn and Phillips 1980; Pathirana and Andersen 1986; Elliott et al. 1989).

Chemical Defenses against Competition and Fouling

Chemical defenses may also be important in spatial competition among coral reef invertebrates and in preventing other species from fouling the surfaces of marine invertebrates. Jackson and Buss (1975) proposed that allelopathy might help maintain high diversity in cryptic coral reef environments where predation and disturbance seem unimportant. They exposed different species of sessile invertebrates (bryozoans, serpulid worms, bivalves, and brachiopods) to homogenates of sponges and ascidians suspected of being allelopathic and found that five of nine sponges and one of two ascidians exhibited species-specific toxic effects. The results provided preliminary evidence that secondary chemicals might play a role in competition; however, they did not demonstrate that the active chemical substances were exuded by the sponges or ascidians. The toxins could be stored within the sponges and ascidians and not released into surrounding seawater.

Natural exudation of secondary metabolites has been proposed as an explanation for bare zones around soft corals and sponges (Coll et al. 1982a; Sullivan et al. 1983; Porter and Targett 1988), infrequent fouling or overgrowth of sponges (Jackson and Buss 1975; Thompson et al. 1985), and reduced larval recruitment within sponge-dominated assemblages (Goodbody 1961; Thompson 1984; Thompson et al. 1985). Thompson (1985) found that the sponge *Aplysina fistularis* released the biologically active metabolites aerothionin (29) and homoaerothionin (30)

29 n = 4
30 n = 5

into seawater at a rate sufficient to affect the behavior and survivorship of several animals capable of harming the sponge in nature. The affected animals included potential fouling organisms and predators such as hydroids, bryozoans, limpets, and sea stars. In a related study, Walker et al. (1985) confirmed that *A. fistularis* exudes aerothionin and homoaerothionin into seawater. Immediately after injury the sponge exuded these metabolites at a rate 10–100 times the normal rate. They also found that sponges with clean surfaces exhibited greater antimicrobial activity than those with fouled surfaces. Aerothionin and homoaerothionin are located in spherulous cells within *A. fistulans.* These cells secrete the compounds into the intercellular matrix and surrounding seawater (Thompson et al. 1983).

Studies with soft corals are among the best demonstrations of competitive interactions mediated by secondary metabolites in the marine environment. Several studies of competitive interactions among soft corals and between soft corals and scleractinian corals have been conducted on the Great Barrier Reef (Coll et al. 1982a; La Barre and Coll 1982; Coll and Sammarco 1983; Sammarco et al. 1983, 1985; La Barre et al. 1986a; Sammarco and Coll 1988). Many soft coral terpenoids cause tissue necrosis, avoidance responses, and growth inhibition in competing species—which might be other soft corals or scleractinian corals. In their review, Coll and Sammarco (1988) pointed out that the effectiveness of soft coral chemical defenses against scleractinian corals is not absolute. Not all soft corals release secondary metabolites into the environment, and not all scleractinians are damaged by soft coral exudates (Sammarco et al. 1985). Similarly, some soft corals are damaged by defense mechanisms of scleractinians (sweeper tentacles, nematocysts, overtopping), while others are not affected (Coll and Sammarco 1988). Soft coral–algal competitive interactions have recently been examined (de Nys et al. 1991). As is true in other ecological interactions, the effects of secondary metabolites in competitive interactions are highly species specific.

Porter and Targett (1988) demonstrated similar allelochemical interactions between the liver sponge (*Plakortis halichondroides*) and the stony coral (*Agaricia lamarcki*) in the Caribbean. The sponge kills the coral on direct contact and also indirectly through waterborne metabolites. Crude organic extracts of *Plakortis* coated on synthetic cellulose sponges tied to the coral caused bleaching of the corals within 24 hours. Control synthetic sponges had no effect. Sponges inhibited the photosynthetic abilities of the symbiotic zooxanthellae in *Agaricia* tissues by reducing the number of zooxanthellae, reducing the maximum net and gross photosynthetic rates, and increasing the nocturnal respiration rate of the coral.

Wahl (1989) discussed antifouling mechanisms of marine organisms, including chemical defenses. Davis et al. (1989) recently reviewed the literature on the inhibition of fouling, settlement, and overgrowth by marine algae and invertebrates. It is clear that few studies have demonstrated that particular metabolites actively exuded by marine invertebrates deter fouling organisms, nor have relevant field studies examining the effects of particular secondary metabolites on the settlement of natural communities of fouling organisms been conducted.

Most marine surfaces are covered with microbial films and diatoms that have been shown to influence (and usually enhance) the settlement of many marine invertebrates (ZoBell and Allen 1935; Mihm et al. 1981; Woollacott 1981; Brancato and Woollacott 1982; Kirchman et al. 1982, reviewed by Davis et al. 1989; Wahl 1989). Many sessile marine invertebrates have very specific chemical settling cues (Williams 1964; Chia and Rice 1978; Burke 1983; Morse and Morse 1984; Pawlik, this volume). Marine natural products may influence settling by affecting the primary microbial films or directly inhibiting or enhancing larval settlement.

Many marine natural products have antimicrobial activities in laboratory assays; however, these tests are usually conducted with terrestrial, not marine, bacteria. Thus, these antimicrobial activities may not be representative of effects on fouling microorganisms (Bakus et al. 1990). Some extracts of tropical sponges showed antimicrobial activity to terrestrial and marine bacteria when tested (McCaffrey and Endean 1985). One example using marine bacteria is a study with didemnenones A (20) and B (21) from *Trididemnum* cf. *cyanophorum*, which significantly inhibited the growth of two *Vibrio* species and the pathogenic marine fungus *Lagenidium callinectes* (Lindquist et al. 1988). In addition, only a few isolated natural products from invertebrates have been examined for antifouling activities. Several metabolites from octocorals were shown to be (1) toxic to diatoms in laboratory assays (Targett et al. 1983; Targett 1985; Bandurraga and Fenical 1985), (2) toxic to marine flagellates (Ciereszko and Guillard 1989), (3) inhibitors of settlement by barnacle larvae (Keifer et al. 1986; Gerhart et al. 1988), or (4) toxic to nudibranch larvae (Hadfield and Ciereszko 1978). In laboratory assays, natural products isolated from southern California sponges inhibited the settlement of bryozoan, polychaete, or abalone larvae, or all three, from the same habitats (Nakatsu et al. 1983; Thompson et al. 1985). Metabolites from the North Carolina sponge *Lissodendoryx isodictyalis* inhibited settlement of barnacle larvae in laboratory assays (Sears et al. 1990); however, active metabolites were not chemically defined. Alkaloids (eudistomins) produced by the ascidian *Eudistoma oli-*

vaceum inhibited settlement of cheilostome bryozoan larvae in laboratory assays (Davis and Wright 1990).

More larval settlement inhibition studies have been conducted with extracts or aqueous homogenates of marine invertebrates than with pure compounds. In general, these studies show that many marine invertebrates contain compounds that are toxic for larvae of some species (Young and Chia 1981; Rittschof et al. 1985, 1986). However, the toxic or inhibitory effects on larvae vary greatly depending on the species of fouling organisms (Dyrynda 1985; Rittschof et al. 1988; Davis and Wright 1989).

Bak and Borsboom (1984) reported that a waterborne substance from the sea anemone *Condylactus gigantea* interferes with the germination of algal spores in the vicinity of the anemone. Similarly, some soft coral diterpenes have been demonstrated to limit the growth of the common alga *Ceramium codii* in culture (Coll et al. 1987). One diterpenoid, 2-epi-sarcophytoxide (31),

31

was isolated from colonies of soft coral (*Lobophytum pauciflorum*) that had been overgrown by algae, while conspecific colonies that had not been overgrown produced two other diterpenoids, 14-hydroxycembra-1,3,7,11-tetraene (32) and 15-hydroxycembra-1,3,7,11-tetraene (33).

32   **33**

Many studies of toxic or antimicrobial activities of isolated marine secondary metabolites are not realistic assessments of the antifouling properties of these compounds under natural conditions. Several questions need to be addressed simultaneously: Are natural products concentrated on the surface of organisms or exuded at sufficient concentrations to deter settlement by fouling orgainsms? Do secondary metabolites inhibit species of larvae that may potentially foul the organism in nature? Even when secondary metabolites show toxicity in laboratory assays, is there evidence that they influence the settling of fouling organisms under natural conditions? These are not easy questions to address. As I have already discussed, compounds may inhibit one fouling organism but not another. A sessile invertebrate may produce some metabolites that inhibit settling by organisms and other metabolites that enhance settling (Standing et al. 1984).

Some field experiments have attempted to test for allelopathic release of chemicals from invertebrates. Bingham and Young (1991) looked at the ability of sponges to deter other organisms from settling nearby. They found that the settlement of most recruiting species was actually greater next to the sponges. Additionally, Hay (1986, this volume) suggests that algal growth and survivorship may be enhanced next to chemically defended algae or invertebrates because these areas are rarely grazed and provide safe sites for palatable seaweeds. Several studies support this hypothesis (Hay 1985, 1986; Littler et al. 1986, 1987; Pfister and Hay 1988). These observations demonstrate the complexity of determining natural antifouling functions for marine secondary metabolites.

Conclusions

As with most areas of marine chemical ecology, we have only begun to scratch the surface of research on marine invertebrates. In many ways, these studies and the questions we can address are much the same as those involving chemically defended plants and herbivorous consumers. Models and predictions involving the evolution of chemical defenses in plants could also be applied to sessile marine invertebrates.

In general, we need to develop more realistic assays to test the deterrent effects of marine secondary metabolites on natural predators, competitors, and fouling organisms. Field assays seem to be particularly useful for examining deterrent activities with regard to communities of predatory fishes, or larvae of potential fouling organisms that may settle on or near sessile invertebrates. We also need to consider that many secondary metabolites may have multiple defensive functions. Some compounds may deter predators, competitors, or pathogenic organisms or all three. Many marine organisms may produce a variety of secondary metabolites because different compounds affect different species of predators and pathogens. Specialists often seem to overcome these chemical defenses. Predatory nudibranchs can even use invertebrate secondary metabolites for their own chemical defenses. Many questions can be asked about the behavior and physiology of these specialists. Finally, variation in the production of secondary metabolites among body parts, among individuals within a species, and among life history stages of marine invertebrates has not been carefully examined. The extent of this chemical variation and its implications for chemical defenses against predators, competitors, and pathogens will certainly be an area of future research.

Acknowledgments   I am grateful to Emmett Duffy, Niels Lindquist, and Steve Nelson for helpful comments on drafts of this chapter. My research program on chemical defenses of marine invertebrates is funded by the National Institutes of Health (GM 38624).

References

Abbott, D.P., and Newberry, A.T. 1980. Urochordata: the tunicates. *In* Intertidal invertebrates of California, ed. R.H. Morris, D.P. Abbott, and E.C. Haderlie, pp. 177–226. Stanford: Stanford University Press.

Amade, P., Pesando, D., and Chevolot, L. 1982. Antimicrobial activities of marine sponges from French Polynesia and Brittany. Mar. Biol. 70:223–228.

Anderson, G.R.V., Ehrlich, A.H., Ehrlich, P.R., Roughgarden, J.D., Russel, B.C., and Talbot, F.H. 1981. The community structure of coral reef fishes. Am. Nat. 117:476–495.

Bak, R.P.M., and Borsboom, J.L.A. 1984. Allelopathic interaction between a reef coelenterate and benthic algae. Oecologia 63:194–198.

Bakus, G.J. 1964. The effects of fish-grazing on invertebrate evolution in shallow tropical waters. Allan Hancock Found. Occas. Pap. 27:1–29.

Bakus, G.J. 1969. Energetics and feeding in shallow marine waters. Int. Rev. Gen. Exp. Zool. 4:275–369.

Bakus, G.J. 1981. Chemical defense mechanisms and fish feeding behavior on the Great Barrier Reef, Australia. Science 211:497–499.

Bakus, G.J., and Abbott, D.P. 1980. Porifera: the sponges. *In* Intertidal invertebrates of California, ed. R.H. Morris, D.P. Abbott, and E.C. Haderlie, pp. 21–39. Stanford: Stanford University Press.

Bakus, G.J., and Green, G. 1974. Toxicity in sponges and holothurians: a geographic pattern. Science 185:951–953.

Bakus, G.J., Schulte, B., Jhu, S., Wright, M., Green, G., and Gomez, P. 1990. Antibiosis and antifouling in marine sponges: laboratory versus field studies. *In* New perspectives in sponge biology, ed. K. Rützler, pp. 102–108. Washington, D.C.: Smithsonian Institution Press.

Bandurraga, M.M., and Fenical, W. 1985. Isolation of the muricins: evidence of a chemical adaptation against fouling in the marine octocoral *Muricea fructicosa* (Gorgonacea). Tetrahedron 41:1057–1065.

Bayer, F.M. 1961. The shallow water octocorals of the West Indian region. The Hague: Martinus Nijhoff.

Bertness, M.D., Garrity, S.D., and Levings, S.C. 1981. Predation pressure and gastropod foraging: a tropical-temperate comparison. Evolution 35:995–1007.

Bingham, B.L., and Young, C.M. Mar. 1991. Influence of sponges on invertebrate recruitment: a field test of allelopathy. Mar. Biol. 109:19–26.

Birkeland, C., and Gregory, B. 1975. Foraging behavior and rates of feeding of the gastropod *Cyphoma gibbosum* (Linnaeus). Nat. Hist. Mus. Los Ang. Cty. Bull. Sci. 20:57–67.

Birkeland, C., and Neudecker, S. 1981. Foraging behavior of two Caribbean chaetodontids: *Chaetodon capistratus* and *C. aculeatus*. Copeia 1981:169–178.

Brancato, M.S., and Woollacott, R.M. 1982. Effect of microbial films on settling of bryozoan larvae. Mar. Biol. 71:551–556.

Burke, R.D. 1983. The induction of metamorphosis of marine invertebrate larvae: stimulus and response. Can. J. Zool. 61:1701–1719.

Carté, B., and Faulkner, D.J. 1983. Defensive metabolites from three nembrothid nudibranchs. J. Org. Chem. 48:2314–2318.

Carté, B., and Faulkner, D.J. 1986. Role of secondary metabolites in feeding associations between a predatory nudibranch, two grazing nudibranchs, and a bryozoan. J. Chem. Ecol. 12:795–804.

Chia, F., and Rice, M.E., eds. 1978. Settlement and metamorphosis of marine invertebrate larvae. New York: Elsevier.

Ciereszko, L.S., and Guillard, R.R.L. 1989. The influence of some cembranolides from gorgonian corals on motility of marine flagellates. J. Exp. Mar. Biol. Ecol. 127:205–210.

Coll, J.C., Bowden, B.F., Tapiolas, D.M., and Dunlap, W.C. 1982a. In situ isolation of allelochemicals released from soft corals (Coelenterata: Octocorallia): a totally submersible sampling device. J. Exp. Mar. Biol. Ecol. 60:293–299.

Coll, J.C., La Barre, S., Sammarco, P.W., Williams, W.T., and Bakus, G.J. 1982b. Chemical defences in soft corals (Coelenterata: Octocorallia) of the Great Barrier Reef: a study of comparative toxicities. Mar. Biol. Prog. Ser. 8:271–278.

Coll, J.C., Price, J.R., Konig, G.M., and Bowden, B.F. 1987. Algal overgrowth of alcyonacean soft corals. Mar. Biol. 96:129–135.

Coll, J.C., and Sammarco, P.W. 1983. Terpenoid toxins of soft corals (Cnidaria: Octocorallia): their nature, toxicity, and ecological significance. Toxicon Suppl. 3:69–72.

Coll, J.C., and Sammarco, P.W. 1988. The role of secondary metabolites in the chemical ecology of marine invertebrates: a meeting ground for biologists and chemists. *In* Proc. Sixth Int. Coral Reef Symp., vol. 1, pp. 167–174.

Coll, J.C., Tapiolas, D.M., Bowden, B.F., Webb, L., and Marsh, H. 1983. Transformation of soft coral (Coelenterata: Octocorallia) terpenes by *Ovula ovum* (Mollusca: Prosobranchia). Mar. Biol. 74:35–40.

Davis, A.R., Targett, N.M., McConnell, O.J., and Young, C.M. 1989. Epibiosis of marine algae and benthic invertebrates: natural products chemistry and other mechanisms inhibiting settlement and overgrowth. *In* Bioorganic marine chemistry, vol. 3, ed. P.J. Scheuer, pp. 85–113. Berlin: Springer-Verlag.

Davis, A.R., and Wright, A.E. 1989. Interspecific differences in fouling of two congeneric ascidians: is surface acidity an effective defense? Mar. Biol. 102:491–497.

Davis, A.R., and Wright, A.E. 1990. Inhibition of larval settlement by natural products from the ascidian *Eudistoma olivaceum* (Van Name). J. Chem. Ecol. 16:1349–1357.

Dayton, P.K. 1989. Interdecadal variation in an Antarctic sponge and its predators from oceanographic climate shifts. Science 245:1484–1486.

de Nys, R., Coll, J.C., and Price, I.R. 1991. Chemically mediated interactions between the red alga *Plocamium hamatum* (Rhodophyta) and the octocoral *Sinularia cruciata* (Alcyonacea). Mar. Biol. 108:315–320.

Duffy, J.E., and Paul, V.J. 1992. Prey nutritional quality and the effectiveness of chemical defenses against tropical reef fishes. Oecologia, in press.

Dyrynda, P.E.J. 1985. Chemical defences and the structure of subtidal epibenthic communities. *In* Proc. Nineteenth Eur. Mar. Biol. Symp., ed. P.E. Gibbs, pp. 411–424. Cambridge: Cambridge University Press.

Elliott, J.K., Ross, D.M., Pathirana, C., Miao, S., Andersen, R.J., Singer, P., Kokke, W.C.M.C., and Ayer, W.A. 1989. Induction of swimming in *Stomphia* (Anthozoa: Actiniaria) by imbricatine, a metabolite of the asteroid *Dermasterias imbricata*. Biol. Bull. 176:73–78.

Fenical, W., and Pawlik, J.R. 1991. Defensive properties of secondary metabolites from the Caribbean gorgonian coral *Erythropodium caribaeorum*. Mar. Ecol. Prog. Ser., in press.

Fishlyn, D.A., and Phillips, D.W. 1980. Chemical camouflaging and behavioral defenses against a predatory seastar by three species of gastropods from the surfgrass *Phyllospadix* community. Biol. Bull 158:34–48.

Gaines, S.D., and Lubchenco, J. 1982. A unified approach to marine plant-herbivore interactions. II. Biogeography. Annu. Rev. Ecol. Syst. 13:111–138.

Gerhart, D.J. 1984. Prostaglandin A$_2$: an agent of chemical defense in the Caribbean gorgonian *Plexaura homomalla*. Mar. Ecol. Prog. Ser. 19:181–187.

Gerhart, D.J. 1986. Gregariousness in the gorgonian-eating gastropod *Cyphoma gibbosum*: tests of several possible causes. Mar. Ecol. Prog. Ser. 31:255–263.

Gerhart, D.J., Rittschof, D., and Mayo, S.W. 1988. Chemical ecology and the search for marine antifoulants: studies of a predator-prey symbiosis. J. Chem. Ecol. 14:1905–1917.

Goodbody, I. 1961. Inhibition of the development of a marine sessile community. Nature 190:282–283.

Goodbody, I., and Gibson, J. 1974. The biology of *Ascidia nigra* (Savigny). V. Survival in populations settled at different times of the year. Biol. Bull. 146:217–237.

Gosliner, T. 1987. Nudibranchs of southern Africa. Monterey, Calif.: Sea Challengers.

Green, G. 1977. Ecology of toxicity in marine sponges. Mar. Biol. 40:207–215.

Hadfield, M.G., and Ciereszko, L.S. 1978. Action of cembranolides derived from octocorals on larvae of the nudibranch *Phestilla sibogae*. *In* Drugs and food from the sea: myth or reality? ed. P.N. Kaul and C.J. Sinderman, pp. 145–150. Norman: University of Oklahoma Press.

Harmelin-Vivien, M.L., and Bouchon-Navaro, Y. 1981. Trophic relationships among chaetodontid fishes in the Gulf of Aqaba (Red Sea). *In* Proc. Fourth Int. Coral Reef Symp., vol. 2, pp. 537–544.

Harmelin-Vivien, M.L., and Bouchon-Navaro, Y. 1983. Feeding diets and significance of coral feeding among chaetodontid fishes in Moorea (French Polynesia). Coral Reefs 2:119–127.

Harvell, C.D., and Fenical, W. 1989. Chemical and structural defenses of Caribbean gorgonians (*Pseudopterogorgia* spp.): intracolony localization of defense. Limnol. Oceanogr. 34:380–387.

Harvell, C.D., Fenical, W., and Greene, C.H. 1988. Chemical and structural defenses of Caribbean gorgonians (*Pseudopterogorgia* spp.). I. Development of an in situ feeding assay. Mar. Ecol. Prog. Ser. 49:287–294.

Harvell, C.D., and Suchanek, T.H. 1987. Partial predation on tropical gorgonians by *Cyphoma gibbosum* (Gastropoda). Mar. Ecol. Prog. Ser. 38:37–44.

Hay, M.E. 1985. Spatial patterns of herbivore impact and their importance in maintaining algal species richness. *In* Proc. Fifth Int. Coral Reef Congr., vol. 4, pp. 29–34.

Hay, M.E. 1986. Associational plant defenses and the maintenance of species diversity: turning competitors into accomplices. Am. Nat. 128:617–641.

Hay, M.E., Duffy, J.E., and Fenical, W. 1988. Seaweed chemical defenses: among-compound and among-herbivore variance. *In* Proc. Sixth Int. Coral Reef Symp., vol. 3, pp. 43–48.

Hay, M.E., and Fenical, W. 1988. Marine plant-herbivore interactions: the ecology of chemical defense. Annu. Rev. Ecol. Syst. 19:111–145.

Hazlett, B.A., and Bach, C.A. 1982. Distribution of the flamingo tongue shell (*Cyphoma gibbosum*) on its gorgonian prey (*Briareum asbestinum*). Mar. Behav. Physiol. 8:305–309.

Hobson, E.S. 1974. Feeding relationships of teleostean fishes on coral reefs in Kona, Hawaii. Fish. Bull. U.S. 72:915–1031.

Ireland, C.M., Roll, D.M., Molinski, T.F., McKee, T.C., Zabriskie, T.M., and Swersey, J.C. 1988. Uniqueness of the marine chemical environment: categories of marine natural products from invertebrates. *In* Biomedical importance of marine organisms, ed. D. Fautin, pp. 41–58. Cal. Acad. Sci. Symp. No. 13. San Francisco: California Academy of Sciences. Mem.

Jackson, J.B.C., and Buss, L. 1975. Allelopathy and spatial competition among coral reef invertebrates. Proc. Natl. Acad. Sci. USA 72:5160–5163.

Keifer, P.A., Rhinehart, K.L., Jr., and Hooper, I.R. 1986. Renillafoulins, antifouling diterpenes from the sea pansy *Renilla reniformis* (Octocorallia). J. Org. Chem. 51:4450–4454.

Kinzie, R.A. 1970. The ecology of the gorgonians (Cnidaria, Octocorallia) of Discovery Bay, Jamaica. Ph.D. dissertation, Yale University, New Haven.

Kirchman, D., Graham, S., Reish, D., and Mitchell, R. 1982. Bacteria induce settlement and metamorphosis of *Janua* (*Dexiospira*) *brasiliensis* Grube (Polychaeta: Spirorbidae). J. Exp. Mar. Biol. Ecol. 56:153–163.

Koehl, M.A.R. 1982. Mechanical design of spicule-reinforced connective tissue. J. Exp. Biol. 98:239–268.

La Barre, S.C., and Coll, J.C. 1982. Movement in soft corals: an interaction between *Nephthea brassica* (Coelenterata: Octocorallia) and *Acropora hyacinthus* (Coelenterata: Scleractinia). Mar. Biol. 72:119–124.

La Barre, S.C., Coll, J.C., and Sammarco, P.W. 1986a. Competitive strategies of soft corals (Coelenterata, Octocorallia). III. Spacing and aggressive interactions between alcyonaceans. Mar. Ecol. Prog. Ser. 28:147–156.

La Barre, S.C., Coll, J.C., and Sammarco, P.W. 1986b. Defensive strategies of soft corals (Coelenterata: Octocorallia) of the Great Barrier Reef. II. The relationship between toxicity and feeding deterrence. Biol. Bull. 171:565–576.

Lasker, H.R. 1985. Prey preferences and browsing pressure of the butterflyfish *Chaetodon capistratus* on Caribbean gorgonians. Mar. Ecol. Prog. Ser. 21:213–220.

Lasker, H.R., and Coffroth, M.A. 1988. Temporal and spatial variability among grazers: variability in the distribution of the gastropod *Cyphoma gibbosum* on octocorals. Mar. Ecol. Prog. Ser. 43:285–295.

Lasker, H.R., Coffroth, M.A., and Fitzgerald, L.M. 1988. Foraging patterns of *Cyphoma gibbosum* on octocorals: the roles of host choice and feeding preference. Biol. Bull. 174:254–266.

Lindquist, N. 1989. Secondary metabolite production and chemical adaptations in the class Ascidiacea. Ph.D. dissertation, University of California, San Diego.

Lindquist, N., and Fenical, W. 1991. New tambjamine class alkaloids from the marine ascidian *Atapozoa* sp. and its nudibranch predators—origin of the tambjamines in *Atapozoa*. Experientia 47:504–506.

Lindquist, N., Fenical, W., Sesin, D.F., Ireland, C.M., Van Duyne, G.D., Forsyth, C.J., and Clardy, J. 1988. Isolation and structure determination of the didemnenones, novel cytotoxic metabolites from tunicates. J. Am. Chem. Soc. 110:1308–1309.

Littler, M.M., Littler, D.S., and Taylor, P.R. 1987. Animal-plant defense associations: effects on the distribution and abundance of tropical reef macrophytes. J. Exp. Mar. Biol. Ecol. 105:107–121.

Littler, M.M., Taylor, P.R., and Littler, D.S. 1986. Plant defense associations in the marine environment. Coral Reefs 5:63–71.

McCaffrey, E.J., and Endean, R. 1985. Antimicrobial activity of tropical and subtropical sponges. Mar. Biol. 89:1–8.

McClintock, J.B. 1987. Investigation of the relationship between invertebrate predation and biochemical composition, energy content, spicule armament and toxicity of benthic sponges at McMurdo Sound, Antarctica. Mar. Biol. 94:479–487.

McClintock, J.B., Heine, J., Slattery, M., and Weston, J. 1991. Biochemical and energetic composition, population biology, and chemical defense of the Antarctic ascidian *Chemidocarpa verrucosa* Lesson. J. Exp. Mar. Biol. Ecol. 147:163–175.

McClintock, J.B., and Janssen, J. 1990. Pteropod abduction as a chemical defence in a pelagic Antarctic amphipod. Nature 346:462–464.

Mebs, D., Weiler, I., and Heinke, H.F. 1985. Bioactive proteins from marine sponges: screening of sponge extracts for hemagglutinating, hemolytic, ichthyotoxic and lethal properties and isolation and characterization of hemagglutinins. Toxicon 23:955–962.

Meylan, A. 1988. Spongivory in hawksbill turtles: a diet of glass. Science 239:393–395.

Mihm, J.W., Banta, W.C., and Loeb, G.I. 1981. Effects of adsorbed organic and primary fouling films on bryozoan settlement. J. Exp. Mar. Biol. Ecol. 54:167–179.

Millar, R.H. 1971. The biology of ascidians. Adv. Mar. Biol. 9:1–100.

Morse, A.N.C., and Morse, D.E. 1984. Recruitment and metamorphosis of *Haliotis* larvae induced by molecules uniquely available at the surfaces of crustose red algae. J. Exp. Mar. Biol. Ecol. 75:191–215.

Myers, R.F. 1983. The comparative ecology of the shallow-water species of *Canthigaster* (Family Tetraodontidae) of Guam. M.S. thesis, University of Guam.

Nakatsu, T., Walker, R.P., Thompson, J.E., and Faulkner, D.J. 1983. Biologically-active sterol sulfates from the marine sponge *Toxadocia zumi*. Experientia 39:759–761.

Olson, R.R. 1983. Ascidian-*Prochloron* symbiosis: the role of larval photoadaptations in midday larval release and settlement. Biol. Bull. 165:221–240.

Parry, D.L. 1984. Chemical properties of the test of ascidians in relation to predation. Mar. Ecol. Prog. Ser. 17:279–282.

Pathirana, C., and Andersen, R.J. 1986. Imbricatine, an unusual benzyltetrahydroisoquinoline alkaloid isolated from the starfish *Dermasterias imbricata*. J. Am. Chem. Soc. 108:8288–8289.

Paul, V.J., Lindquist, N., and Fenical, W. 1990. Chemical defenses of the tropical ascidian *Atapozoa* sp. and its nudibranch predators *Nembrotha* spp. Mar. Ecol. Prog. Ser. 59:109–118.

Paul, V.J., Wylie, C., and Sanger, H. 1988. Chemical defenses of tropical seaweeds:

effects against different coral-reef herbivorous fishes. *In* Proc. Sixth Int. Coral Reef Symp., vol. 3, pp. 73–78.

Pawlik, J.R., Burch, M.T., and Fenical, W. 1987. Patterns of chemical defense among Caribbean gorgonian corals: a preliminary survey. J. Exp. Mar. Biol. Ecol. 108:55–66.

Pawlik, J.R., and Fenical, W. 1989. A re-evaluation of the ichthyodeterrent role of prostaglandins in the Caribbean gorgonian coral *Plexaura homomalla*. Mar. Ecol. Prog. Ser. 52:95–98.

Pawlik, J.R., Kernan, M.R., Molinski, T.F., Harper, M.K., and Faulkner, D.J. 1988. Defensive chemicals of the Spanish dancer nudibranch *Hexabranchus sanguineus* and its egg ribbons: macrolides derived from a sponge diet. J. Exp. Mar. Biol. Ecol. 119:99–109.

Pfister, C.A., and Hay, M.E. 1988. Associational plant refuges: convergent patterns in marine and terrestrial communities result from differing mechanisms. Oecologia 77:118–129.

Porter, J.W., and Targett, N.M. 1988. Allelochemical interactions between sponges and corals. Biol. Bull. 175:230–239.

Randall, J.E. 1967. Food habits of reef fishes of the West Indies. Stud. Trop. Oceanogr. 5:665–847.

Randall, J.E. 1974. The effects of fishes on coral reefs. *In* Proc. Second Int. Coral Reef Symp., vol. 1, pp. 159–166.

Randall, J.E., and Hartman, W.D. 1968. Sponge-feeding fishes of the West Indies. Mar. Biol. 1:216–225.

Rittschof, D., Hooper, I.R., Branscomb, E.S., and Costlow, J.D. 1985. Inhibition of barnacle settlement and behavior by natural products from whip corals, *Leptogorgia virgulata* (Lamarck, 1815). J. Chem. Ecol. 11:551–563.

Rittschof, D., Hooper, I.R., and Costlow, J.D. 1986. Barnacle settlement inhibitors from sea pansies, *Renilla reniformis*. Bull. Mar. Sci. 39:376–382.

Rittschof, D., Hooper, I.R., and Costlow, J.D. 1988. Settlement inhibition of marine invertebrate larvae: comparison of sensitivities of bryozoan and barnacle larvae. *In* Marine biodeterioration. Advanced techniques applicable to the Indian Ocean, ed. M.F. Thompson, R. Sarojini, and R. Nagabhushanam, pp. 151–163. New Delhi: Oxford and IBH Publishing.

Rogers, S.D., and Paul, V.J. 1991. Chemical defenses of three *Glossodoris* nudibranchs and their dietary *Hyrtios* sponges. Mar. Ecol. Prog. Ser., in press.

Sammarco, P.W., and Coll, J.C. 1988. The chemical ecology of alcyonarian corals. *In* Bioorganic marine chemistry, vol. 2, ed. P.J. Scheuer, pp. 89–116. Berlin: Springer-Verlag.

Sammarco, P.W., Coll, J.C., and La Barre, S.C. 1985. Competitive strategies of soft corals (Coelenterata: Octocorallia). II. Variable defensive response and susceptibility to scleractinian corals. J. Exp. Mar. Biol. Ecol. 91:199–215.

Sammarco, P.W., Coll, J.C., La Barre, S.C., and Willis, B. 1983. Competitive strategies of soft corals (Coelenterata: Octocorallia): Allelopathic effects on selected scleractinian corals. Coral Reefs 1:173–178.

Sammarco, P., La Barre, S., and Coll, J.C. 1987. Ichthyotoxicity and morphology of soft corals. Oecologia 74:93–102.

Schneider, W.P., Bundy, G.L., Lincoln, F.H., Daniels, E.G., and Pike, J.E. 1977. Isolation and chemical conversions of prostaglandins from *Plexaura homomalla*: preparation of prostaglandin E_2, prostaglandin F_2, and their 5,6-*trans* isomers. J. Am. Chem. Soc. 99:1222–1232.

Sears, M.A., Gerhart, D.J., and Rittschof, D. 1990. Antifouling agents from marine sponge *Lissodendoryx isodictyalis* Carter. J. Chem. Ecol. 16:791–799.

Shanks, A.L., and Graham, W.M. 1988. Chemical defense in a scyphomedusa. Mar. Ecol. Prog. Ser. 45:81–86.

Standing, J., Hooper, I.R., and Costlow, J.D. 1984. Inhibition and induction of barnacle settlement by natural products present in octocorals. J. Chem. Ecol. 10:823–834.

Steneck, R.S. 1986. The ecology of coralline algal crusts: convergent patterns and adaptive strategies. Annu. Rev. Ecol. Syst. 17:273–303.

Steneck, R.S. 1988. Herbivory on coral reefs: a synthesis. *In* Proc. Sixth Int. Coral Reef Symp., vol. 1, pp. 37–49.

Stoecker, D. 1978. Resistance of a tunicate to fouling. Biol. Bull. 155:615–626.

Stoecker, D. 1980a. Distribution of acid and vanadium in *Rhopalaea birkelandi* Tokioka. J. Exp. Mar. Biol. Ecol. 48:277–281.

Stoecker, D. 1980b. Relationships between chemical defense and ecology in benthic ascidians. Mar. Ecol. Prog. Ser. 3:257–265.

Stoecker, D. 1980c. Chemical defenses of ascidians against predators. Ecology 61:1327–1334.

Sullivan, B., Faulkner, D.J., and Webb, L. 1983. Siphonodictidine, a metabolite of the burrowing sponge *Siphonodictyon* sp. that inhibits coral growth. Science 221:1175–1176.

Targett, N.M. 1985. Allelochemistry in marine organisms: chemical fouling and antifouling strategies. *In* Marine biodeterioration. Advanced techniques applicable to the Indian Ocean, ed. M.F. Thompson, R. Sarojini, and R. Nagabhushanam, pp. 609–617, New Delhi: Oxford and IBH Publishing.

Targett, N.M., Bishop, S.S., McConnell, O.J., and Yoder, J.A. 1983. Antifouling agents against the benthic marine diatom *Navicula salinicola*: homarine from the gorgonian *Leptogorgia virgulata* and *L. setacea* and analogs. J. Chem. Ecol. 9:817–829.

Thompson, J.E. 1984. Chemical ecology and the structure of sponge dominated assemblages. Ph.D. dissertation, University of California, San Diego.

Thompson, J.E. 1985. Exudations of biologically-active metabolites in a sponge (*Aplysina fistularis*). I. Biological evidence. Mar. Biol. 88:23–26.

Thompson, J.E., Barrow, K.D., and Faulkner, D.J. 1983. Localization of two brominated metabolites, aerothionin and homoaerothionin, in spherulous cells of the marine sponge *Aplysina fistularis* (= *Verongia thiona*). Acta Zoologica 64:199–210.

Thompson, J.E., Murphy P.T., Bergquist, P.R., and Evans, E.A. 1987. Environmentally induced variation in diterpene composition of the marine sponge *Rhopaloeides odorabile*. Biochem. Syst. Ecol. 15:595–606.

Thompson, J.E., Walker, R.P., and Faulkner, D.J. 1985. Screening and bioassays for biologically-active substances from forty marine sponge species from San Diego, California, USA. Mar. Biol. 88:11–21.

Tursch, B., and Tursch, A. 1982. The soft coral community on a sheltered reef quadrat at Laing Island (Papua New Guinea). Mar. Biol. 68:321–332.

Van Alstyne, K.L., and Paul, V.J. 1992. Chemical and structural antipredator defenses in the sea fan *Gorgonia ventalina*: effects against generalist and specialist predators. Coral Reefs, in press.

Van Alstyne, K.L., Wylie, C.R., and Paul, V.J. 1992. Intracolony variation in the production of morphological defenses in three species of soft corals, *Sinularia* spp.: effects against generalist carnivorous fishes. Biol. Bull., in press.

van Duyl, F.C., Bak, R.P.M., and Sybesma, J. 1981. The ecology of the tropical compound ascidian *Trididemnum solidum*. I. Reproductive strategy and larval behavior. Mar. Ecol. Prog. Ser. 6:35–42.

Vermeij, G.J. 1978. Biogeography and adaptation. Cambridge: Harvard University Press.

Vermeij, G.J. 1987. Evolution and escalation. Princeton: Princeton University Press.

Vreeland, H.V., and Lasker, H.R. 1989. Selective feeding of the polychaete *Hermodice carunculata* Pallas on Caribbean gorgonians. J. Exp. Mar. Biol. Ecol. 129:265–277.

Wahl, M. 1989. Marine epibiosis. I. Fouling and antifouling: some basic aspects. Mar. Ecol. Prog. Ser. 58:175–189.

Walker, R.P., Thompson, J.E., and Faulkner, D.J. 1985. Exudation of biologically-active metabolites in the sponge *Aplysina fistularis*. II. Chemical evidence. Mar. Biol. 88:27–32.

Willan, R.C., and Coleman, N. 1984. Nudibranchs of Australasia. Australian Marine Photographic Index. Sydney, Australia.

Williams, G.B. 1964. The effects of extracts of *Fucus serratus* in promoting the settlement of larvae of *Spirorbis borealis* (Polychaeta). J. Mar. Biol. Assoc. U.K. 44:397–414.

Woollacott, R.M. 1981. Association of bacteria with bryozoan larvae. Mar. Biol. 65:155–158.

Wylie, C.R., and Paul, V.J. 1989. Chemical defenses in three species of *Sinularia* (Coelenterata, Alcyonacea): effects against generalist predators and the butterflyfish *Chaetodon unimaculatus* Bloch. J. Exp. Mar. Biol. Ecol. 129:141–160.

Young, C.M., and Bingham, B.L. 1987. Chemical defense and aposematic coloration in larvae of the ascidian *Ecteinascidia turbinata*. Mar. Biol. 96:539–544.

Young, C.M., and Chia, F.S. 1981. Laboratory evidence for delay of larval settlement in response to a dominant competitor. Int. J. Invertebr. Reprod. 3:221–226.

ZoBell, C.E., and Allen, E.C. 1935. The significance of marine bacteria in the fouling of submerged surfaces. J. Bacteriol. 29:239–251.

Chapter 6
Induction of Marine Invertebrate Larval Settlement: Evidence for Chemical Cues

JOSEPH R. PAWLIK

The invertebrates of the marine benthos include representatives of all the major phyla of animals, and some, such as the echinoderms, live exclusively in this domain. Approximately 80% of marine invertebrates, or roughly 90,000 species, produce microscopic larvae that develop in the plankton (Thorson 1964). These larvae, which have morphologies and diets completely unlike those of their parents, may drift great distances before contacting a suitable substratum and metamorphosing into their adult form. In this chapter, I review evidence for environmental chemical signals that initiate the transition from planktonic larva to benthic adult. The information presented here is largely excerpted from a review of the chemical ecology of marine invertebrate settlement (Pawlik 1992). It is presented in this volume for the sake of completeness and to benefit readers who might otherwise not encounter it in the literature on marine biology.

Until the latter half of this century, the predominant opinion held by marine biologists was that invertebrate larvae metamorphosed in the water column and sank to the bottom, having little ability to affect their distribution; hence, the site of settlement was randomly determined (e.g., Petersen 1913; Colman 1933). This view gave way as considerable evidence accumulated that larvae respond to various physical and biological factors, and, moreover, they delay settlement until a suitable substratum is contacted (reviewed in Thorson 1966; Meadows and Campbell 1972; Scheltema 1974; Burke 1983; Crisp 1984). More recently, the processes that control the recruitment of invertebrate larvae have sparked considerable interest among ecologists (Connell 1985; Gaines and Roughgarden 1985; Sutherland 1990; Menge 1991), with some concomitant debate over the importance of active substratum selection versus passive deposition in determining subsequent patterns of abundance (reviewed in Butman 1987). There is ample experimental evidence, however, that differential larval settlement largely predicts patterns of benthic recruitment, particularly on hard substrata (R. R. Strathmann et al. 1981; Keough 1983; Watanabe 1984; Bushek 1988; Raimondi 1991); and active substratum selection has been demonstrated in laboratory experiments under defined flow conditions (Butman et al. 1988; Pawlik et al. 1991).

Invertebrate larvae are exposed to a multitude of environmental factors during their lives in the plankton and at the time of settlement, and undoubtedly they respond to many stimuli in the course of substratum selection. Larvae are known to respond to physical factors, including light, gravity, hydrostatic pressure, salinity, temperature, and water flow (Thorson 1964; Crisp 1984; Sulkin 1984; Young and Chia 1987; Pawlik et al. 1991), and substratum-associated factors such as contour, texture, thermal capacity, and sediment characteristics (Ryland 1974; Raimondi 1988a; Walters and Wethey 1991). But the biological and chemical nature of the substratum has proved to have the greatest influence on larval settlement in both laboratory and field studies (R. R. Strathmann and Branscomb 1979; Mihm et al. 1981; LeTourneux and Bourget 1988; Raimondi 1988b).

The ample literature on the biology of marine invertebrate larvae is filled with anecdotal and experimental accounts of settlement induced by specific substrata; most of these accounts emphasize the importance of chemical signals in mediating larval behavior (reviews in Meadows and Campbell 1972; Chia and Rice 1978; Crisp 1984; Butman 1987; Chia 1989). In this chapter I review the various categories of substratum-specific settlement exhibited by marine invertebrates and then focus attention on studies that have advanced beyond the implication of a settlement cue to the full or partial characterization of a chemical inducer. I do not discuss endogenous chemical signals (hormones) that presumably control the intricate metamorphic processes of most marine invertebrates, about which little is known, nor do I address the inhibition of settlement (antifouling) by chemical means, a topic covered elsewhere (Pawlik 1992; Paul, this volume).

Site-Specific Settlement

Aggregative Settlement

The formation of monospecific colonies and aggregations is very common among marine invertebrates. Colonies made up of individuals that are genetically similar or the same (clones) are formed by the asexual division of an initial settler (as in the growth of a coral head from a single polyp) or by the settlement of direct-developing or short-term pelagic larvae near their mother (as with some sponges, ascidians, and bryozoans). The mobile adults of some species aggregate temporarily for the purpose of breeding (Pennington 1985). For most species, aggregations of genetically unrelated individuals are formed by the settlement of planktonic larvae on or near adult conspecifics. This last condition is particularly prevalent among hard-bottom, sessile intertidal organisms, including barnacles (e.g., Knight-Jones 1953; Raimondi 1991), bivalves (Bayne 1969; McGrath et al. 1988), and polychaetes (Wilson 1968; Scheltema et al. 1981). Gregarious settlement has been reported in at least 35 invertebrate species representing eight phyla; in 18 of these there was evidence of a chemical inducer of settlement (Burke 1986).

Gregariousness has many advantages (Crisp 1979; Pawlik and Faulkner 1988). Larvae that settle on or near adult conspecifics have chosen a habitat more likely to support postlarval growth than one chosen indiscriminately. Juveniles may derive additional benefits from the presence of adult conspecifics; for example, juvenile sand dollars that recruit to beds of adults encounter less predation from tanaid crustaceans, which are displaced by the sediment-reworking activities of the adult sand dollars (Highsmith 1982). Adult invertebrates also derive reproductive benefits from aggregation. Proximity increases fertilization success for both internally fertilizing and freely spawning species (Crisp 1979; Pennington 1985). Moreover, individuals in aggregations may live longer than solitary individuals (Wilson 1974) and thereby benefit from greater fecundity over the course of a longer adult life span.

Associative Settlement

The term *associative settlement* was first used by Crisp (1974) to describe the enhanced or specific settlement of one species on another. Inasmuch as associative settlement results in heterospecific organisms living in close proximity, it can be subdivided into several categories on the basis of the nutritional relationship of the adult organisms. Nonparasitic associations, or symbioses, between organisms are variously defined as mutualistic (reciprocally advantageous), commensalistic (one party benefits, the other is neither helped nor harmed), inquilinistic (association for protection), epibiotic (association for substratum), and phoretic (association for transport) (Zann 1980). Many invertebrates involved in these symbioses settle as larvae onto their hosts. In most cases, chemical cues are thought to be responsible for this specificity.

Mutualism and commensalism are common in marine communities, particularly in the tropics. Among crustaceans, a variety of shrimps and crabs (representative genera include *Periclimenes, Pontonia, Pinnotheres, Petrolisthes*, and *Trapezia*) are associated with specific host cnidarians, echinoderms, and molluscs (Zann 1980; Stevens 1990). Some barnacles are extremely specialized in their substratum requirements, settling specifically on sponges, gorgonians, hard corals, turtles, sea snakes, or marine mammals (reviewed in Crisp 1974; Lewis 1978; Zann 1980).

Epiphytic associations are common among several species of marine invertebrates that have planktonic larvae. Encrusting red algae promote the settlement of various species of corals (Sebens 1983; D. E. Morse et al. 1988), polychaetes (Gee 1964), and echinoderms (Barker 1977; Rowley 1989), while some bivalves prefer filamentous red algae (Eyster and Pechenik 1987). Brown algae are the preferred substrata of hydroids, spirorbid polychaetes, and bryozoans (Nishihira 1968; Scheltema 1974). Various species of spirorbids settle specifically on red algae of the genera *Corallina* or *Lithothamnion*, sea grasses of the genera *Posidonia, Thallasia*, or *Zostera*, or on the shells of crustaceans (Crisp 1974; Dirnberger 1990).

Most invertebrate phyla include parasitic groups; some are made up entirely of parasites. Little is known about how the dispersive stages of marine parasites find their hosts, but it is likely that chemical signals are involved. Molluscan parasites include pyramidellaceans, which are ectoparasites on other molluscs and polychaetes, and coralliophilids, snails that live inside or next to the corals they parasitize (Hadfield 1976).

Among the crustaceans are several parasitic groups with pelagic larval stages, including isopods, copepods, and cirripedes. Perhaps the most specialized of these are the rhizocephalan barnacles, which primarily parasitize other crustaceans (Høeg and Lützen 1985). Female rhizocephalans form dendritic processes in the bodies of their hosts; when mature, each produces a reproductive sac external to the host from which larvae are released. Cyprid larvae are either male or female. The female larva must locate a suitable host to infect, usually a specific crustacean species, while the male larva must settle on the reproductive sac of a virginal female. After injecting spermatogenic cells into the female sac, the male cyprid dies. Considering the unlikelihood of a random encounter between a female cyprid and its host, or, more so, between a male cyprid and a virginal female reproductive sac, the involvement of chemical cues at settlement seems a necessity. In addition, the cyprid larvae of some rhizocephalans lack thoracic appendages and cannot swim (Pawlik 1987), making the chance discovery of a virgin adult female by a male larva even more improbable.

Herbivorous and predatory marine invertebrates with specific food requirements and pelagic larvae are generally thought to settle on or near their prey. Settlement cues are probably involved whenever the prey organisms have highly restricted distributions. Among molluscs, several opisthobranch groups have very narrow food requirements: specific cnidarian prey for most aeolid nudibranchs, sponges for dorid nudibranchs, siphonaceous green algae for ascoglossans, and red algae or cyanophytes for aplysiids (Switzer-Dunlap 1978; Faulkner and Ghiselin 1983; Hadfield and Miller 1987). Some chitons and gastropod molluscs settle specifically on encrusting red algae and feed on the crusts after metamorphosis (Barnes and Gonor 1973; A. N. C. Morse and Morse 1984a).

Settlement on Microbial Films

The growth of microorganisms on hard substrata or sediments has long been recognized as a prerequisite for the settlement of some invertebrates (Zobell and Allen 1935; Gray 1974; Scheltema 1974). Clean surfaces exposed to seawater go through a succession of changes, beginning with the formation of a primary film of organic material and advancing to the development of a complex microbial community (Mitchell and Kirchman 1984). Although microorganisms promote the settlement of many species, including hydroids (e.g., Müller 1973), polychaetes (Kirchman et al. 1982a), bivalves (Weiner et al. 1985), bryozoans (Mihm et al. 1981), barnacles (LeTourneux and Bourget 1988), and echinoderms (Cameron and Hinegardner 1974), microbial films may inhibit the settlement of others (Maki et al. 1988, 1989).

The Chemical Nature of Settlement Inducers

Considering the many examples in which marine invertebrate larval settlement has been demonstrated to be highly substratum specific, it is surprising that naturally occurring settlement inducers have been isolated and identified for only a few species. This lack of knowledge stems largely from the difficulties associated with obtaining competent larvae (i.e., larvae that are developmentally ready to settle). Only rarely can they be harvested directly from the plankton (e.g., Rice 1988); generally, field-caught larvae are difficult to identify and their densities are too low to provide sufficient numbers for replicate experiments. Laboratory culture of larvae is an obvious alternative. It is relatively easy to obtain the gametes of some invertebrates, but other species are notoriously difficult to spawn or maintain narrow reproductive seasons (M. Strathmann 1987). Once sufficient numbers of newly hatched larvae have been procured, they must be maintained in culture for the time necessary for them to become competent to settle. This period is usually short for lecithotrophic (nonfeeding) larvae, but most pelagic larvae are planktotrophic and must be fed for several weeks before competence is attained. Despite the difficulties, experiments have been performed on the responses of several species to chemical cues that have been purified and characterized to various degrees.

Isolated and Identified Settlement Inducers

The structures of substratum-derived, naturally occurring compounds that stimulate settlement are known for only four species of marine invertebrates: the hydroid *Coryne uchidai*, the echiuran *Bonellia viridis*, the bivalve *Pecten maximus*, and two subspecies of the polychaete *Phragmatopoma lapidosa*. The information that follows regarding compounds that affect the first three species has been taken primarily from the chemical literature; biological evidence to support their putative function is somewhat equivocal.

Coryne uchidai   Nishihira (1968) studied the settlement specificity of the hydroid *Coryne uchidai* on brown algae of the family Sargassaceae. When placed in assay dishes containing 20 1-mm^2 pieces of *Sargassum thunbergii*, *S. confusum*, or *S. tortile*, larvae of *C. uchidai* stopped swimming immediately and began crawling. In contrast, in dishes without algae or in dishes containing an equal quantity of the green alga *Ulva pertusa*, larvae gradually ceased swimming and began crawling over the course of a day. The abrupt drop to the bottom in the presence of algal exudates resulted from the cessation of larval ciliary activity. Most of the larvae formed polyps within two to three days in dishes containing *S. confusum* or *S. tortile*, within three to four days in dishes containing *S. thunbergii* and *U. pertusa*, and within three to six days in control dishes (no algae).

Boiled aqueous extracts of *Sargassum tortile* similarly caused larvae to cease swimming immediately, with metamorphosis occurring within two days. Extracts of *U. pertusa* had no effect on larvae beyond the gradual settlement observed in control dishes. Extracts of two other algae, *Dictyopteris divaricata* and *Symphyocladia latiuscula* (neither of which belong to the family Sargassaceae), caused immediate cessation of larval swimming but little or no metamorphosis. An extract of the latter alga killed larvae within one day. Choice experiments were performed with extracts incorporated into agar blocks, but the results were ambiguous because the aqueous extracts leached out of the blocks and into the seawater in the assay dishes.

Fractionation of hexane extracts of dried *Sargassum tortile* led to the isolation of several diterpenoid chromanols (Kato et al. 1975). Two of the most abundant chromanols were identified as δ-tocotrienol (1) and its epoxide (2).

Only limited settlement assay results were provided. Compounds were dissolved in a drop of ethanol and added to 20 ml of seawater containing 10 larvae. After 72 hours, δ-tocotrienol at 37.5 μg/ml seawater induced 3 larvae to metamorphose, but the remaining 7 died after settlement. The epoxide induced all 10 larvae to metamorphose within 72 hours at both 18.8 and 75 μg/ml seawater. None of the larvae in control dishes (ethanol alone) completed metamorphosis within 72 hours.

Unfortunately, no further information is available on the settlement responses of *C. uchidai* to algal metabolites. It is unclear whether chromanols are the only inductive compounds present in *S. tortile*, and whether species of *Sargassum* produce these compounds to the exclusion of nonpreferred algae. Inasmuch as the compounds described are lipophilic, it is unlikely that they are perceived in solution; yet a water-soluble inducer was indicated by experiments with aqueous extracts of algae. It is also unclear whether the chromanols are elaborated by the algae in such a way that larvae would encounter them in nature.

Bonellia viridis During the early part of this century, Baltzer (reviewed in Pilger 1978; Jaccarini et al. 1983) discovered that sexually undifferentiated larvae of *Bonellia viridis* were stimulated to settle and metamorphose into nonfeeding dwarf males after contact with the proboscis of an adult female. Larvae that failed to encounter an adult female metamorphosed and developed into females themselves. Male development appeared to result primarily from the inhibitory effects of an unknown factor on female development. Aqueous extracts of the female proboscis and intestine induced metamorphosis into males at concentrations of 1 part dried tissue to 6000–9000 parts seawater. Herbst (in Pilger 1978; Jaccarini et al. 1983) found that similar effects could be triggered by altering the ionic composition or pH of the seawater to which larvae were exposed. Baltzer proposed that the masculinizing factor was bonellin, a green integumentary pigment that had been isolated from adult females in 1875 by Sorby (for a history of chemical investigations of *B. viridis*, see Agius et al. 1979). A century later, the structure of bonellin was described as an uncomplexed, alkylated chlorin (3) (Pelter et al. 1978).

3

Whereas bonellin was the predominant isolate of the proboscis of *B. viridis*, amino acid conjugates of the compound were present in the body wall of the animal (Cariello et al. 1978; Ballantine et al. 1980). It was suggested that these conjugates of bonellin were stored or scavenged in the body of *B. viridis* and elaborated as bonellin in the proboscis in order to affect larval settlement. Agius (1979), however, found the masculinizing factor in aqueous extracts of proboscides and body tissues, and potent activity in the pigmented body secretion. Larvae were attracted to the proboscides of female worms and absorbed the green pigment from the proboscis at the site of their attachment (Agius et al. 1979). Larval assays performed with purified bonellin resulted in ambiguous data: at 1 ppm, the compound induced 31.2% of the larvae to differentiate into males, as opposed to 99% in the presence of an adult female and 8.5% in control seawater. Strangely, 0.5 and 0.2 ppm bonellin induced 35.8% and 14.4% masculinization, respectively, but 0.01 ppm bonellin induced 44.5% of the larvae to develop into males.

Results of a more rigorous study were presented by Jaccarini et al. (1983), who concluded that the effects of bonellin on sex determination were inconsistent. Significantly higher levels of masculinization were induced in larvae exposed to 10^{-6} M bonellin, but while purified bonellin induced 28% of the larvae to develop into males, the female body secretion triggered 96% masculinization. Moreover, in three of five experiments with purified bonellin, there was no enhanced masculinization over controls. There had been a suggestion that bonellin induced masculinization by a photodynamic effect: the compound is toxic to a wide range of organisms when assayed in the presence of light (Agius et al. 1979; De Nicola Giudici 1984). But comparative larval assays of purified bonellin in light and darkness resulted in no significant effects on controls (Jaccarini et al. 1983). Therefore there is no unequivocal evidence linking bonellin to the masculinizing properties of the body secretion of *B. viridis*. Enhancement of larval settlement on, or attraction of larvae to, the proboscides of female worms has yet to be experimentally assessed.

Pecten maximus   Yvin et al. (1985) reported that aqueous ethanol extracts of the red alga *Delesseria sanguinea* stimulated settlement of the bivalve scallop *Pecten maximus*. The active component was partitioned into ether, purified by high-performance liquid chromatography (HPLC), and identified as jacaranone (4),

$$O$$

OH

COOCH$_3$

4

a compound previously isolated from *Jacaranda caucana*, a terrestrial vascular plant. Jacaranone stimulated maximum settlement of *P. maximus* at 0.5 mg/l ($\sim 3 \times 10^{-6}$ M), with increasing levels of larval mortality at higher concentrations (Cochard et al. 1989). This response does not appear to be particularly relevant to the biology of *P. maximus*, however, because the species is not known to settle with any degree of specificity on *D. sanguinea*; its recruitment patterns are relatively indiscriminate (Cochard et al. 1989).

Phragmatopoma   Marine polychaete worms of the family Sabellariidae live in tubes constructed of cemented grains of sand. Some species are gregarious and form colonies and reefs of amassed sand tubes. These colonies are entirely dependent on the recruitment of planktonic larvae for reef maintenance and growth (Pawlik and Faulkner 1988). Wilson (1968, 1974) studied the larval settlement behavior of several sabellariids from British waters, in particular *Sabellaria alveolata*. Settlement was stimulated on contact with adult tubes, tube remnants, or the mucoid tubes of juvenile worms. Factors such as surface contour and roughness, sediment type, water motion, and the presence of surface microorganisms had only a minor influence on larval behavior. The settlement-inducing capacity of the tubes was insoluble in water and unaffected by drying but was destroyed by cold concentrated acid. Wilson concluded that a chemical cue in the tube cement triggered larval settlement in a fashion similar to that proposed for barnacle larvae by Knight-Jones (1953; see Gregarious Settlement, below).

Larvae of *Phragmatopoma californica*, a gregarious sabellariid from the coast of California, also chose the sand tubes of adult conspecifics over other substrata (Jensen and Morse 1984). Sequential extraction of the tube sand of *P. californica* in a series of organic solvents diminished its capacity to induce larval settlement (Pawlik 1986). The inductive activity was retained in the organic extracts of natural tube sand. An active fraction was isolated from the extracts by HPLC, and nuclear magnetic resonance spectrometry and gas chromatographic analysis of the fraction revealed that it consisted of a mixture of free fatty acids (FFAs) ranging from 14 to 22 carbons in length. Extracts of worm-free tube sand from reefs formed by *P. californica* contained concentrations of FFAs sufficient to induce larval settlement (Pawlik 1986).

The FFA fraction isolated from the tube sand of *P. californica* contained predominantly eicosapentaenoic (20:5), palmitic (16:0), and palmitoleic acids (16:1). (In the shorthand notation for FFAs, the number of carbon atoms in the molecule precedes the colon, and the number of double bonds follows.) Of the 9 FFAs that contributed 3% or more to the active fraction, only palmitoleic, linoleic (18:2), arachidonic (20:4), and eicosapentaenoic acids (5–8, respectively) induced larval settlement.

5    $CH_3(CH_2)_5CH=CH(CH_2)_7COOH$

6    $CH_3(CH_2)_4CH=CHCH_2CH=CH(CH_2)_7COOH$

7    $CH_3(CH_2)_4(CH=CHCH_2)_3CH=CH(CH_2)_3COOH$

8    $CH_3CH_2(CH=CHCH_2)_4CH=CH(CH_2)_3COOH$

In further assays of an additional 28 FFAs of variable carbon chain length and unsaturation, larval response was stereospecific, with maximum settlement in response to palmitoleic, linolenic (18:3), eicosapentaenoic, and docosahexaenoic (22:6) acids (Pawlik and Faulkner 1986). Palmitelaidic acid, the *trans* isomer of highly active palmitoleic acid, was ineffective at inducing larval settlement. The capacity to stimulate settlement was linked to molecular shape, which is determined both by the number of carbon atoms and the number of *cis* double bonds in the acyl chain. For example, although palmitoleic acid (16:1) was a potent inducer of larval settlement, oleic acid (18:1) was not, as a result of its greater molecular length. Linoleic (18:2) and linolenic acids (18:3) were active, however, because the additional *cis* double bonds act to twist and shorten these molecules to an overall shape similar to that of palmitoleic acid. The induction of larval settlement by FFAs was also dependent on the presence of a free carboxyl group. Modification of the carboxyl terminus of the FFA molecule by esterification or reduction resulted in the loss of inductive activity (Pawlik and Faulkner 1986). Therefore, larval response was dependent on the presence of at least one *cis* double bond in the molecule, conservation of molecular shape with increasing acyl chain length by addition of *cis* double bonds, and the presence of a free carboxyl group. This high degree of stereochemical specificity was likened to that described in studies of chemoreception in terrestrial insects (Pawlik and Faulkner 1986, 1988; see Comparisons with Terrestrial Insects, below).

Larval settlement responses of *Phragmatopoma lapidosa*, a gregarious sabellariid from the tropical western Atlantic, were very similar to those of *P. californica* (Pawlik 1988b). The inductive capacity of the tube sand was lost on extraction and the activity was retained in the extracts. Again, a suite of FFAs was isolated as the active component. The same FFAs that stimulated larval settlement of *P. californica* did so for *P. lapidosa*, and they were isolated from the natural tube sand of both species at about the same concentrations. The similarities did not end there: in addition to having identical larvae and adults, the two species were completely interfertile in reciprocal fertilization experiments. The hybrid larvae of both crosses developed and metamorphosed normally, prompting the synonymization of the two species: *Phragmatopoma lapidosa lapidosa* for the western Atlantic subspecies, and *P. l. californica* for the eastern Pacific subspecies.

Larvae of *Sabellaria alveolata*, a reef-building sabellariid from European waters, did not respond to the same chemical signals as responded to by *P. l. californica* (Pawlik 1988a). In reciprocal assays, settlement of both species occurred to a greater extent on conspecific tube sand than on heterospecific tube sand. Extraction of the tube sand of *S. alveolata* diminished its capacity to trigger settlement of conspecific larvae, but the capacity was not transferred to the organic extracts, and an inducer was not isolated or identified. Furthermore, the FFAs that elicited settlement of *P. l. californica* and *P. l. lapidosa* either were not effective at inducing settlement of *S. alveolata* or actually inhibited settlement. Larvae of nongregarious species from the Caribbean, *S. floridensis*, and the eastern Pacific, *S. cementarium*, similarly did not respond to FFAs (Pawlik 1988b; Pawlik and Chia 1991). FFAs were present in the natural tube sand of *S. alveolata* at less than one-tenth the concentration found in natural tube sand of *P. l. californica* and *P. l. lapidosa*, suggesting that adults of the two subspecies of *Phragmatopoma* produce the FFAs that induce conspecific settlement. Pawlik concluded that settlement of different genera of gregarious sabellariids is under the control of different chemical signals. Interspecific differences in larval responses to FFAs further suggested that a specific mechanism is responsible for the perception of FFAs by larvae of the two subspecies of *P. lapidosa*.

Jensen and Morse, who also studied *P. l. californica*, arrived at different conclusions regarding the induction of settlement of this species (Jensen and Morse 1984, 1990; Yool et al. 1986; Jensen et al. 1990). Building on a hypothesis put forth by Wilson (1968), they suggested that some component of quinone-tanned proteins, specifically, an unidentified, cross-linked residue of the amino acid L-β-3,4–dihydroxyphenylalanine (L-DOPA, 9), present in the tube cement of adult worms was responsible for inducing settlement. Larval responses to solutions of L-DOPA were weak at best (Jensen and Morse 1984; Pawlik 1990), but settlement occurred readily in response to the cresol-derived, lipophilic compound 2,6-di-tert-butyl-methylphenol, also known as butylated hydroxytoluene (BHT, 10) (Jensen and Morse 1990).

9

10

BHT effected settlement of *P. l. californica* when adsorbed to surfaces in both laboratory and field experiments. Jensen and Morse (1990) proposed that BHT mimics the activity of the unknown, naturally occurring L-DOPA residue from tube cement (but see Pawlik 1990). In addition, the authors questioned whether FFAs function as a natural cue, as proposed by Pawlik (1986; 1988b), suggesting instead that FFAs induce settlement in a nonspecific manner, possibly by operating on the larval nervous system or parallel to the natural inducer (Jensen and Morse 1990; Jensen et al. 1990). In support of their contention: (1) they were unable to detect FFAs on glass beads used by adult worms to make tubes (the natural inducer); (2) freeze-drying and stirring reduced the inductive activity of the natural inducer but not the activity of glass beads coated with FFAs; (3) induction by FFAs was temperature dependent, while induction by the natural inducer was not; and, (4) induction by the natural inducer was taxon specific, but induction by FFAs was not (Jensen et al. 1990).

Do FFAs function as natural settlement cues for larvae of *Phragmatopoma*? Further study is necessary to answer this question. Replication of experiments detailed in Jensen et al. (1990) has failed to confirm that freeze-drying or stirring decreases the inductive activity of the natural inducer as compared with substrata coated with FFAs, or that induction by FFAs is temperature dependent (points 2 and 3 above; Pawlik, unpubl. data). Moreover, the contention that larval response to FFAs is nonspecific (point 4 above) was supported by Jensen et al. (1990) with data from highly variable assays of abalone larvae (*Haliotis rufescens*), while the high degree of specificity of larval response within the Sabellariidae was ignored (Pawlik 1988a, 1988b). But, clearly, if FFAs are absent from inductive, uncontaminated tube sand, the naturally occurring cue must lie elsewhere. Jensen et al. (1990) may be correct in suggesting that the natural tube sand used by Pawlik (1986, 1988b) was contaminated with organic material containing FFAs (possibly oocytes that stuck to sand grains as gravid adult females were removed from their tubes). But if FFAs are not the natural inducers of settlement, then it is unclear why larval responses to these compounds are restricted to the genus *Phragmatopoma* within the polychaete family Sabellariidae, and why these compounds occur at high concentrations in the natural tube sand of species that respond to them, but not in the tube sand of species that do not (Pawlik 1986, 1988a, 1988b; Pawlik and Chia 1991).

Larval settlement experiments performed in laboratory flumes have several major advantages over those performed in still water; in particular, the ability of larvae to select substrata in flow can be assessed (see Butman 1987; Butman et al. 1988). *P. l. californica* has proven to be a very useful subject in experiments with flumes because (1) the larvae are large enough to be easily seen; (2) they undergo settlement rapidly; (3) metamorphosis results in major morphological changes, permitting the separation of larvae and metamorphosed juveniles in fixed samples; and (4) settlement is highly specific, occurring only on tube sand or sand treated with inductive compounds. Pawlik et al. (1991) conducted flume experiments at two flow regimes in which larvae of *P. l. californica* were offered a choice of five treatment substrata in a five-by-five Latin-square array. In both flows, larvae settled preferentially on the two substrata that had induced settlement in still-water assays (tube sand and sand treated with palmitoleic acid). Surprisingly, delivery of larvae to the array was greater in fast flow, because larvae tended to move off the bottom in slow flow. Therefore behavior may be important in the settlement process at two levels for *P. l. californica*: larvae respond first to flow conditions and then, as they sample the substratum, to chemical cues.

Partially Purified Inducers

Studies of several invertebrate species described below have proceeded toward characterizing the chemical induction of settlement but have not yet identified the stimulatory compounds. After field observations and laboratory assays indicated preferential settlement of larvae onto a specific substratum, research generally advanced along two lines: (1) various physical and chemical treatments were used in an attempt to destroy the stimulatory capacity, and (2) an effort was made to isolate the inductive factor (often with the use of dialysis tubing) or to transfer it onto an otherwise inactive substratum. If the stimulatory capacity was isolated or transferred, more rigorous chemical separation techniques were often employed.

Gregarious Settlement Partially purified inducers of gregarious settlement have been described for several species. Because of their commercial importance, oyster larvae have been the subjects of considerable interest. Crisp (1967) discovered that chemical removal of the organic outer layers of the shell of *Crassostrea virginica* reduced the settlement of conspecific larvae on that substratum, while aqueous extracts of adult animals enhanced settlement. High levels of settlement occurred on tiles treated with lyophilized aqueous extracts of whole oysters or material from aqueous extracts of whole oysters that had been partitioned into diethyl ether (Keck et al. 1971). Hidu (1969) found that settlement of *C. virginica* was stimulated by the water held between the valves of living adults. Larval settlement was promoted by a protein-containing fraction purified from this water (Veitch and Hidu 1971). The protein component had a molecular mass greater than 10 kilodaltons and contained iodinated amino acids. A protease-labile fraction isolated from an oyster tissue extract also enhanced the settlement of *Ostrea edulis* (Bayne 1969). Acetazolamide, an inhibitor of the enzyme carbonic anhydrase, promoted settlement of the New Zealand oyster *Ostrea lutaria*, but its mechanism of action remains unexplained (Nielsen 1973). More recent work on *Crassostrea gigas* has yielded settlement inducers of bacterial origin (Coon et al. 1988; Fitt et al. 1990; see Microbial Films, and Inorganic Compounds as Inducers, below).

Laboratory and field experiments have demonstrated that the sessile gastropod molluscs *Crepidula fornicata* and *C. plana* preferentially settle near adult congeners (McGee and Targett 1989). Larvae of *C. fornicata* exhibited the highest levels of metamorphosis in response to water conditioned by adult conspecifics, but *C. plana* settled in response to water conditioned by either species or by the hermit crab *Pagurus pollicaris*, which inhabits mollusc shells that are often encrusted with adult *Crepidula*. Metamorphosis-inducing activity in seawater conditioned by adult *C. plana* passed through both 10- and 5-kilodalton membrane filters and was retained on a reverse-phase chromatography column (McGee and Targett 1989).

Chemical substances that promote gregarious settlement have been partially purified for the echiuran *Urechis* caupo, the sipunculan *Golfingia misakiana*, and the sand dollars *Dendraster excentricus* and *Echinarachnius parma*. Larvae of *U. caupo* settled rapidly in response to sediments from adult burrows or sediments that had been exposed to the epidermis of an adult worm (Suer and Phillips 1983). The worm-derived factor triggered settlement only when adsorbed onto sediment. It was soluble in seawater and passed through dialysis membrane (3.5–14 kilodaltons), and it was heat labile (>80°C) but stable at ambient seawater temperatures for several days.

Larvae of *Golfingia misakiana* settled in response to a low-molecular-mass (<500 daltons), heat-labile factor present in seawater conditioned by the presence of adult worms (Rice 1988). Larvae of *G. misakiana* did not respond to water conditioned by adults of two other species of sipunculans. In addition to the water-soluble adult factor, sediment covered by a microbial film was required for larvae to begin metamorphosis.

Among echinoderms, chemical induction of larval settlement has been demonstrated most clearly for two species of sand dollars. Highsmith (1982) determined that larvae of *Dendraster excentricus* preferentially settled and metamorphosed on sand from beds of adult conspecifics. The responses of larvae to sand in dialysis tubing and to sand treated with proteolytic enzymes suggested that a small peptide (<10 kilodaltons) was involved. These results were confirmed by Burke (1984), who isolated fractions (by gel-permeation chromatography and HPLC) from extracts of sand from beds of *D. excentricus* that triggered settlement at 10^{-6}–10^{-5} M. Again, a peptide was indicated as the active component, based on a positive reaction using the Lowry method for protein determination and loss of activity on treatment with proteases. Pearce and Scheibling (1990) demonstrated a similar larval response to conspecifics for the sand dollar *Echinarachnius parma*. Sand could be conditioned by the presence of adults (and thereby rendered capable of inducing high levels of metamorphosis) in the dark and after treatment with antibiotics, suggesting that the cue was derived from conspecifics rather than microflora. The water-soluble inductive factor was destroyed by heating and diffused through dialysis tubing with a pore size of 1 kilodalton.

Svane et al. (1987; Havenhand and Svane 1989) studied the effects of aqueous extracts of adult tissues on larvae of the ascidians *Ascidia mentula* and *Ascidiella scabra* and suggested that larval responses to chemical cues in the adult ascidian tunic may lead to gregarious settlement. *Ascidia mentula* showed aggregated recruitment in the field. Larvae of *Ascidia mentula* placed in the middle of a seawater-filled tube that was sealed at one end with the tunic of a living conspecific adult were preferentially distributed near the tunic after 10 minutes. Embryos of both species were treated with extracts during late development and through hatching, and the percentage of tadpole larvae that had resorbed their tails was scored (metamorphosis irrespective of attachment to the substratum). Conspecific adult extracts enhanced metamorphosis of larvae of *Ascidia mentula* but stimulated metamorphosis of *Ascidiella scabra* before they had hatched. Greater inductive activity was associated with the tunic than with the internal tissues of the adult ascidians. Because the extracts did not trigger other components of normal larval settlement (activation of anterior papillae), however, it is unclear how the extracted factors function under natural circumstances.

Investigations of the chemical basis for the gregarious settlement of barnacle larvae have a long history, and the topic has been thoroughly reviewed elsewhere (Crisp 1984; Gabbott and Larman 1987). Most barnacle species liberate feeding nauplius larvae that molt through successive stages. In the last larval molt, each nauplius is transformed into a nonfeeding cyprid larva whose sole purpose is to find a suitable site for settlement. Knight-Jones first described gregarious settlement by *Elminius modestus, Semibalanus balanoides* (= *Balanus balanoides*), and *Balanus crenatus* (Knight-Jones and Stevenson 1950; Knight-Jones 1953), although subsequent work focused primarily on *S. balanoides* (Crisp and Meadows 1963; Larman et al. 1982). The chemical factor responsible for settlement was highly refractory to physical and chemical treatment, was perceived by cyprid larvae only on contact with factor-treated surfaces, and was present in extracts of several barnacle species, other invertebrates, and a fish (Knight-Jones 1953; Crisp and Meadows 1963; Larman and Gabbott 1975). The factor was identified as arthropodin, a proteinaceous component of arthropod cuticles (Crisp and Meadows 1963).

The settlement-inducing substance was further purified and characterized by Larman and Gabbott (1975; Larman et al. 1982; Larman 1984; Gabbott and Larman 1987). Protein precipitates of boiled extracts of adult *S. balanoides* were separated by electrophoresis to yield two fractions, both containing protein and carbohydrate (>50 kilodaltons), that induced barnacle settlement. Molecules exhibiting similar electrophoretic properties, and similar effectiveness at inducing settlement, were isolated from extracts of other invertebrates and a fish. Exhaustive analyses of boiled and unboiled extracts of *S. balanoides* led to the conclusion that the settlement factor was one (or many) of several closely related acidic proteins, homologous with those described from studies of the cuticles of insects and crustaceans (arthropodins) and with amino acid compositions similar to that of actin. Proteins of this class are sticky; in particular, they adhere well to other proteins. More force is required to remove reversibly attached cyprid larvae (as opposed to those that have begun metamorphosis and cemented themselves for permanent attachment) from extract-treated surfaces than from untreated surfaces or surfaces treated with other proteins (Yule and Crisp 1983; Yule and Walker 1984). Moreover, reversibly attached cyprids left behind proteinaceous "footprints" of their own making, which then might stimulate other cyprids to settle (Yule and Walker 1987). Crisp and Meadows (1963) proposed a mechanism by which cyprid larvae detect the settlement factor that relied solely on the stickiness of the inductive proteins (Crisp 1984; Gabbott and Larman 1987). In this instance, settlement was theorized to result from a physical property of the chemical cue (adhesion) rather than from receptor-mediated larval perception, an idea supported by the observed settlement of barnacles on slicks of oil and organometallic compounds (see below). Surprisingly, Crisp (1990) recently found that cyprids of *Balanus amphitrite* settle more readily on conspecific arthropodin than on arthropodins of four other barnacle species, with correspondingly less settlement occurring on the more distantly related species. This is not the expected result if the effects of arthropodins are purely physical. Bourget and colleagues intensively studied the settlement of *S. balanoides* on the Atlantic coast of Canada and concluded that in addition to physical cues, larvae respond to conspecifics (Chabot and Bourget 1988) and to films of microalgae (LeTourneux and Bourget 1988), depending on the spatial scale examined. Mucus has also been observed to affect barnacle settlement (L. E. Johnson and Strathmann 1989). Depending on its source, the mucus either enhanced or inhibited settlement, suggesting that chemical perception rather than stickiness alone may be involved.

Associative Settlement   Partially purified inducers of associative settlement have been described for species in several phyla. Experimental surfaces treated with extracts of fucoid brown algae elicit the settlement of epibiotic bryozoans, bivalves, and spirorbid polychaetes. Larvae of the bryozoan *Alcyonidium polyoum*, an epibiont on *Fucus serratus*, were induced to settle on surfaces treated with aqueous extracts of the alga (Crisp and Williams 1960). Extracts of two other fucoids, *Fucus vesiculosus* and *Ascophyllum nodosum*, also stimulated settlement. Similar responses were observed for larvae of another bryozoan, *Flustrellidra hispida* (Crisp and Williams 1960). Kiseleva (1966) noted that settlement of the bivalve *Brachyodontes lineatus* was stimulated by aqueous extracts of *Cystoseira barbata*. Extracts of *Fucus serratus* similarly promoted settlement of the epibiotic polychaete *Spirorbis borealis* (Williams 1964). Gee (1964) studied another spirorbid, *S. rupestris*, which settled with a high degree of specificity on the crustose coralline red alga *Lithothamnion polymorphum*. Experimental plates treated with aqueous extracts of *L. polymorphum* stimulated high levels of settlement, and the active factor passed through dialysis tubing with an average pore diameter of 24 Å.

Shipworms are bivalve molluscs that bore into wood. Harington (1921) showed that larvae of the shipworm *Teredo norvegica* aggregated around the dried residue of alcohol or ether extracts of wood, the opening of a capillary tube containing seawater saturated with these extracts, or an aqueous extract of sawdust. Aqueous extracts of wood in the form of bogwater (dissolved humic substances, or *Gelbstoff*) also induced crawling behavior in larvae of the shipworms *Teredo navalis* and *Bankia gouldi* (Culliney 1972).

Larvae of the red abalone *Haliotis rufescens* preferentially settle on crustose red algae of the genera *Lithothamnion*, *Lithophyllum*, and *Hildenbrandia* (D. E. Morse et al. 1980c), although gregarious settlement onto the mucus of adult conspecifics has also been reported for this species (Slattery 1987), and abalone are commercially settled on plates covered with benthic microalgae and bacteria (Hahn 1989). The settlement of *H. rufescens* was first attributed to the presence of γ-aminobutyric acid (GABA, 11)

$$(CH_2)_3 - COOH$$
$$|$$
$$NH_2$$

11

molecules "covalently linked" to proteins, and to phycoerythrobilin in the tissues of the algae (D. E. Morse et al. 1979). After further research, however, the settlement factor was limited to a macromolecular fraction isolated from several species of red algae and cyanobacteria but detectable only on the surface of encrusting red algae (A. N. C. Morse and Morse 1984a, 1984b). Treatment of the fraction with proteases or separation by gel-filtration chromatography resulted in the isolation of a group of small (640–1250 daltons), peptide-containing inducer molecules (A. N. C. Morse and Morse 1984a; A. N. C. Morse et al. 1984). The responses of larvae to neuroactive compounds such as GABA are further discussed below and have been reviewed by Hahn (1989) and Pawlik (1990).

Nadeau et al. (1989) reported that larvae of the California sea hare, *Aplysia californica*, did not attain competence to metamorphose when raised in artificial seawater or in natural seawater during the winter months in Woods Hole, Massachusetts. Larvae developed normally when raised in natural seawater during the summer at Woods Hole or during the winter at Hopkins Marine Station, California. This finding suggests that an exogenous factor, absent during the winter in natural seawater off Woods Hole, is required for larval maturation. The factor was inactivated by heating and was retained on 30–100-kilodalton ultrafiltration membranes, but further attempts to isolate it have been unsuccessful. Nadeau et al. (1989) also found that exudates and extracts of the red macroalgae preferentially eaten by juvenile and adult sea hares promoted larval maturation and subsequent metamorphosis in artificial seawater. They proposed that the source of the factors that induce both competence and metamorphosis is red macroalgae and that the same compound controls both processes. Metamorphosis of *A. californica* can occur on a variety of red, green, and brown intertidal macroalgae (Pawlik 1989), however, so it seems unlikely that the putative factor(s) are restricted to red macroalgae.

Hadfield and colleagues have studied the larval settlement of the nudibranch mollusc *Phestilla sibogae*, an obligate predator of hard corals of the genus *Porites* (Hadfield 1977, 1978, 1984; Hadfield and Scheuer 1985; Hadfield and Pennington 1990). Larvae of *P. sibogae* settled in response to a waterborne inducer released from *Porites compressa* or *Porites lobata*, or in response to seawater containing an aqueous coral extract. The inducer was partially purified and characterized as a small polar molecule (200–500 daltons) with broad temperature and pH stability (0–100°C, pH 1–10). The compound triggered settlement at concentrations as low as a few parts per billion (estimated at $\sim 10^{-10}$ M). Larvae exposed to the inducer prior to becoming competent did not metamorphose once they had fully matured. This habituation was reversed if competent larvae were placed in clean seawater before reexposure to the inducer (Hadfield and Scheuer 1985). Despite intensive efforts, the inducer has not yet been identified (Hadfield and Pennington 1990).

Another nudibranch mollusc, *Eubranchus doriae*, settles preferentially on its hydroid prey, *Kirchenpaueria pinnata* (Bahamondes-Rohas and Dherbomez 1990). Fractions of aqueous extracts of the hydroid, prepared by ultrafiltration, induced metamorphosis of the nudibranch larvae. Various sugars dissolved in seawater (10^{-4} M) also induced metamorphosis, provided that the hydroxyl groups attached to carbons 3 and 4 were in the *cis* position (this category includes galactosamine and hexoses such as D-talose and D-galactose). Affinity chromatography of the inductive fraction indicated the presence of galactosidic residues (Bahamondes-Rohas and Dherbomez 1990).

Preliminary data were gathered on the chemical stimulation of settlement of the barnacle *Membranobalanus orcutti*, an obligate sponge commensal (Pawlik, unpublished data). The barnacle settles on and grows into the surface of two sponges in southern California; only one of these, *Spheciospongia confoederata*, is abundant in shallow waters (Jones 1978). Sponge specimens were collected and freeze-dried, and 2-cm-diameter disks of the sponge surface tissue were punched out with a cork borer. Fifty cyprids of *M. orcutti* were added to 700 ml of seawater containing a rehydrated sponge disk or the equivalent surface area of living sponge, and the number of metamorphosed juveniles was recorded after 48 hours. Experiments were performed with three replicates. Larvae readily settled on rehydrated sponge disks; 65 ± 2% (mean ± SD) metamorphosed within two days, as compared with 39 ± 13% metamorphosis in response to living sponge. Removal of lipid-soluble constituents of the sponge disks did not have a great effect on settlement: 41 ± 12% metamorphosed on disks that had been sequentially extracted in organic solvents (hexane, diethyl ether, ethyl acetate, and methanol) prior to rehydration. Boiling sponge disks in fresh water for long periods reduced their capacity to stimulate settlement: disks boiled for 5 minutes induced 37 ± 12% metamorphosis, whereas disks boiled for 30 minutes induced only 21 ± 9%. Immersion of sponge disks in a 37% formaldehyde solution for 1 hour (followed by a 24-hour seawater rinse) destroyed their capacity to stimulate settlement; 0.6 ± 1% metamorphosis occurred on these disks, although larvae used in this assay readily settled when subsequently exposed to untreated sponge disks. Similarly, disks treated with 1 mg/ml of a nonspecific protease (fungal pronase, Sigma P5147) for 12 hours in seawater at room temperature (followed by a 24-hour seawater rinse) induced only 5.3 ± 9% metamorphosis. These results suggest that the chemical settlement cue for *M. orcutti* is a relatively refractory protein associated with the surface tissue of the host sponge.

Microbial Films Chemical factors produced by microorganisms that elicit the settlement of cnidarians, polychaetes, molluscs, and echinoderms have been partially characterized. Müller (1973) described the settlement of planulae of *Hydractinia echinata*, a hydrozoan, in response to "leakage products" from marine gram-negative bacteria. The active factor was identified as a polar lipid that could be partitioned from a cell-free "leakage solution" into chloroform and was unstable and nondialyzable.

Planulae of the scyphozoan *Cassiopea andromeda* settled in response to a 1–100-kilodalton inducer present in the culture medium of a marine bacterium, *Vibrio* sp. (Neumann 1979). Larvae settled after exposure to various peptides, large proteins, glycoproteins, and cholera toxin, but the relationship of these compounds to the bacterially produced inducer is unknown (Fitt and Hoffmann 1985; Fitt et al. 1987; see Induction by Bioactive Compounds, below).

Larvae of the spirorbid polychaete *Janua brasiliensis* settled preferentially on microbial films cultured from the surface of the green alga

Ulva lobata (Kirchman et al. 1982a). Glucose or the lectin concanavalin A (a protein that binds carbohydrate moieties) blocked settlement (Kirchman et al. 1982b). A mechanism of substratum recognition was proposed whereby lectins produced by the larvae bind to specific extracellular bacterial polysaccharides (Maki and Mitchell 1985).

Bonar, Coon, Fitt, Weiner, and their colleagues have produced a body of work on the induction of settlement of oyster larvae (*Crassostrea*) by bacterial products and ammonia. Weiner et al. (1985) isolated a gram-negative, pigment-forming bacterium, designated LST, from oysters and oyster-holding tanks that enhanced the settlement of *Crassostrea virginica*. It was hypothesized that films of LST produce L-dihydroxyphenylalanine (L-DOPA), melanin precursors, and melanin itself, which stimulate larval settlement (Bonar et al. 1985). Coon et al. (1985) discovered that L-DOPA induces settlement behavior of *C. gigas*, thus supporting this hypothesis. Subsequent research revealed, however, that L-DOPA is converted into the neurotransmitter dopamine inside the larva and is not likely to stimulate settlement behavior under natural conditions (Coon and Bonar 1987). A distinction was made between "settlement behavior" (larval foot extension beyond the shell margin) and "metamorphosis" (loss of velum, growth of shell and gill) of *C. gigas*, with separate cues hypothesized for the onset of each (Bonar et al. 1990; Coon et al. 1990a); L-DOPA, for example, induced settlement behavior but not metamorphosis (but see Bonar et al. 1990: table 2). Most recently, the induction of settlement behavior has been attributed to unknown dissolved chemical inducers in the supernatants of cultures of the bacteria *Alteromonas colwelliana* and *Vibrio cholerae* (Fitt et al. 1990). Following size-exclusion chromatography, the inductive activity was retained in a fraction with a molecular mass less than or equal to 300 daltons. Low levels of activity were also detected in the medium used to culture the bacteria (Fitt et al. 1990). Previously, the active component had been identified as dissolved ammonia gas (Coon et al. 1988; Bonar et al. 1990), but this claim has not been repeated (Fitt et al. 1990). Instead, ammonia was proposed as a separate inducer of settlement behavior (Coon et al. 1990b; see Inorganic Compounds as Inducers, below). Metamorphosis of oyster larvae (as opposed to settlement behavior) is thought to be induced by unknown, substratum-associated cues of bacterial origin (Fitt et al. 1990).

Extending the work of Scheltema (1961), Levantine and Bonar (1986) partially isolated a water-soluble factor from sediment that induces metamorphosis of the mud snail *Ilyanassa obsoleta* (= *Nassarius obsoletus*). Ultrafiltration and molecular-sieve chromatography indicated a molecular mass less than 1000 daltons. Analysis of active fractions suggested a high carbohydrate, low protein content.

Cameron and Hinegardner (1974) determined that competent larvae of the sea urchins *Lytechinus pictus* and *Arbacia punctulata* settled in response to a bacterial film or to seawater that had been incubated with particulate material from aquarium filters or sediment from the bottom of seawater storage barrels. The seawater-borne factor was nonvolatile, was removed by adsorption onto charcoal, and had a molecular mass less than 5 kilodaltons.

Induction by Bioactive Compounds

In contrast to the few systems in which the naturally occurring inducers of settlement have been identified, a profusion of bioactive compounds and neuropharmacological agents are known to have varying effects on mature larvae, ranging from normal settlement and metamorphosis to abnormal metamorphosis and death (reviewed in Pawlik 1990).

Neurotransmitters and their derivatives that affect larval responses include choline and succinylcholine chloride (gastropods: Hadfield 1978; polychaetes: Pawlik 1990), DOPA and the catecholamines dopamine, norepinephrine, and epinephrine (bivalves: Cooper 1982; Coon et al. 1985; gastropods: Pires and Hadfield 1991; polychaetes: Jensen and Morse 1990; Pawlik 1990), and GABA (gastropods: D. E. Morse et al. 1979; echinoderms: Pearce and Scheibling 1990). Bound "neurotransmitter mimetics" have been suggested as inducers of settlement and metamorphosis in a variety of invertebrate larvae (D. E. Morse 1985), but criticisms of this theory have been advanced (Pawlik 1990). In the cases in which naturally occurring settlement cues have been isolated, the compounds are unrelated to neurotransmitters; for example, diterpenes for the cnidarian *Coryne uchidai* (Kato et al. 1975) and free fatty acids for two subspecies of the polychaete *Phragmatopoma lapidosa* (Pawlik 1986, 1988b). Neurotransmitters such as DOPA and GABA are water-soluble amino acids; the capacity of invertebrate larvae to actively transport amino acids into their bodies has been clearly established (e.g., Jaeckle and Manahan 1989). It seems most likely that these neuroactive compounds stimulate larval responses by influencing the larval nervous system internally rather than by acting on an epithelial chemoreceptor (Hirata and Hadfield 1986; Coon et al. 1990a).

Other bioactive substances have been assessed for their effects on invertebrate larvae. Stimulatory compounds include those that alter transmembrane ion transport, such as ouabain (hydroid: Müller 1973) and picrotoxin (gastropods: D. E. Morse et al. 1980b; barnacles: Rittschof et al. 1986), and those that affect the intracellular concentrations of cyclic AMP, such as cholera toxin (scyphozoans: Fitt et al. 1987) and isobutylmethylxanthine (gastropods: Baxter and Morse 1987). The responses of larvae to these drugs have been used to formulate complex intracellular signal transduction mechanisms controlling metamorphic activation (Baxter and Morse 1987). But behavioral assays of whole larvae have not been specific, either in the application of the drug or in the assessment of the response, and the validity of models of the molecular pathways controlling settlement and metamorphosis has been called into question (Pawlik 1990).

Inorganic Compounds as Inducers

Some of the earliest research on marine invertebrate larvae concerned the effects of metal salts on settlement and metamorphosis (oysters: Pytherch 1931; Korringa 1940). Copper, iron, and zinc salts have been used to accelerate ascidian and bryozoan settlement (see review in Lynch 1961), although the observed effects have generally been attributed to the toxicity of these compounds.

Elevated concentrations of monovalent cations in seawater induce settlement of many invertebrate species, most probably by affecting the electrical potential across larval cell membranes. Müller and co-workers found that pulsed exposure of larvae of the hydrozoan *Hydractinia echinata* to seawater containing excess Cs^+, Rb^+, Li^+, or K^+ induced metamorphosis in a dose-dependent fashion (Spindler and Müller 1972; Müller and Buchal 1973; for earlier work on ionic effects, see Lynch 1961). Ouabain, a cardiac glycoside that blocks active transport of Na^+ and K^+, inhibited metamorphosis in response to Cs^+, Rb^+, and Li^+, but not to K^+, suggesting that the increase in monovalent cations affects the Na^+, K^+ transport system. Elevated concentrations of K^+ have subsequently been found to induce metamorphosis by the molluscs *Haliotis rufescens*, *Phestilla sibogae*, *Astrea undosa*, *Crepidula fornicata*, and *Adalaria proxima* (Baloun and Morse 1984; Yool et al. 1986; Pechenik and Heyman 1987; Todd et al. 1991), and the polychaete *Phragmatopoma lapidosa californica* (Yool et al. 1986). It seems unlikely, however, that invertebrate larvae encounter elevated concentrations of monovalent cations under natural conditions.

Two inorganic gases dissolved in seawater have more recently been reported to stimulate settlement of invertebrate larvae: hydrogen sulfide (H_2S) and ammonia (NH_3). Cuomo (1985) demonstrated enhanced levels of settlement by larvae of a sediment-dwelling polychaete, *Capitella* sp. I, in the presence of sulfide, with optimal settlement occurring in the 1.0–0.1 mM range. Larvae settled in response to sulfide whether sediment was present or not. Cuomo (1985) suggested that this specific response to sulfide would explain the recruitment of these worms to organically rich sediments. Dubilier (1988) further investigated the settlement of *Capitella* and concluded that the apparent enhancement of settlement in the presence of sulfide was a toxic effect (larvae ceased swimming and lay on the bottom in an apparently anesthetized state), not the response of larvae to a specific settlement cue. She determined that larvae of *Capitella* settled and metamorphosed within minutes in response to organic-rich sediments in the absence of sulfide, whereas addition of sulfide resulted in delayed settlement. Sulfide in the absence of sediment enhanced settlement, but the response required 12–24 hours and resulted in abnormal settlement behavior. Moreover, a similar response was produced by performing the assay in water without H_2S, but depleted of oxygen. Larvae of *Capitella* spp. I and II preferentially settled in organically rich sediments in choice experiments performed in a laboratory flume under hydrodynamic conditions similar to those encountered by larvae in nature (Butman et al. 1988; Grassle and Butman 1989); sulfide levels were likely negligible under these experimental conditions.

Geochemical cues such as sulfide might be important settlement stimuli for the pelagic larvae of hydrothermal vent invertebrates in the deep sea (Lutz et al. 1984; VanDover et al. 1988). Vent organisms rely on the oxidation of sulfide as their ultimate energy source, and vent communities are tightly clustered around hydrothermal apertures that discharge sulfide and a wide variety of other inorganic compounds (Coale et al. 1991). The vents are transitory, and the ability to locate new vent sites is of considerable importance for the larvae of organisms adapted to live there.

Ammonia gas (NH_3) dissolved in seawater causes larvae of the oyster *Crassostrea gigas* to begin substratum exploration ("settlement behavior" *sensu* Coon et al. 1988; Bonar et al. 1990; Coon et al. 1990a; Fitt et al. 1990); an unknown, bacterially derived, surface-associated cue is required for the subsequent onset of metamorphosis (Fitt et al. 1990). Solutions of NH_4Cl and $(NH_4)_2SO_4$ containing as little as 100 μM NH_3 induced larval foot extension, but the response was believed to be the result of increased intracellular pH rather than a specific response to ammonia (Coon et al. 1988, 1990b). Weak bases, such as methylamine and trimethylamine, also induced similar behavior. Ammonia was initially identified as the active agent isolated from bacterial supernatants (Coon et al. 1988; Bonar et al. 1990), a claim that has not been repeated (Fitt et al. 1990). Although ammonia occurs in the water column under natural conditions at concentrations less than an order of magnitude lower than those required for the induction of substratum exploration in oyster larvae, Coon et al. (1990b) suggested that it may play a role in oyster settlement.

Hydrogen peroxide (H_2O_2) at 10^{-4} M in seawater causes the loss of the velum (swimming organ) in larvae of the nudibranch mollusc *Phestilla sibogae* (Pires and Hadfield 1991). Velar loss occurs as part of the normal process of metamorphosis of gastropod veliger larvae; in this case, the response is probably a toxic one. Invertebrate larval responses to oxidized solutions of L-DOPA and catecholamines are often very different from responses to freshly prepared, unoxidized solutions (e.g., Pawlik 1990), an effect that may be attributable to the production of H_2O_2 as these compounds oxidize (Pires and Hadfield 1991).

Induction by Petroleum Products and Organic Solvents

Holland and co-workers (Holland et al. 1984) discovered that oil extracted from Blackstone oil shale contains a factor that enhances barnacle settlement without deleterious side effects. Settlement of *Semibalanus balanoides* was greater on panels of oil shale than on slate panels in both laboratory and field experiments. Moreover, treatment of shale panels with dichloromethane resulted in even higher levels of settlement, presumably resulting from the mobilization of lipophilic inducers from the kerogen matrix of the shale (Huxley et al. 1984). Smith and Hackney (1989) also found that crude oil or a mixture of gasoline and engine oil spread onto clamshells promoted the settlement of the barnacles *Balanus improvisus* and *B. eburneus*, but petroleum treatment inhibited settlement of the oyster *Crassostrea virginica*.

Hill and Holland (1985) fractionated extracts of oil shale and reported enhanced settlement of *Semibalanus balanoides* and *Elminius modestus* in response to an adsorbed layer of a fraction containing metalloporphyrins. The hydrocarbon and asphaltene fractions inhibited settlement. Thin-layer chromatographic separation of the metalloporphyrin fraction yielded three active bands, two of which were identified by ultraviolet spectrometry as nickel- and vanadium-chelated porphyrins. Settlement was also enhanced by commercially available protoporphyrin IX dimethyl ester when chelated with nickel, vanadium, ferrous, or magnesium ions, but unchelated porphyrin (acid or free base) did not enhance settlement. Maximum enhancement was observed at 0.5–1.0 g metalloporphyrin per square meter, depending on the valence of the chelated metal ion. Metalloporphyrins have been hypothesized to stimulate barnacle settlement in much the same way as arthropodin: the compounds are sticky and presumably bind the proteins associated with the cyprid attachment disk (Hill and Holland 1985; see Gregarious Settlement, above; and Chemosensory Organs, below).

Common organic solvents at high concentrations elicited settlement of the nudibranch *Phestilla sibogae* (Pennington and Hadfield 1989). Five alcohols (including ethanol and methanol), ethanolamine, acetonitrile, acetone, dichloromethane, and toluene were effective at inducing settlement, but ethylene glycol, dimethyl sulfoxide, benzene, and hexane were not. A maximum of 65% settlement occurred in response to ethanol at 0.1 M concentrations after 3–5 days; lethal concentrations of ethanol exceeded 0.75 M.

Chemoreception and Settlement

In general, naturally occurring chemical inducers of settlement are tactually perceived by marine invertebrate larvae; that is, contact with the substratum is required for recognition of the cue. This has been demonstrated repeatedly and across phylogenetic lines for cnidarians (Donaldson 1974; D. E. Morse et al. 1988), polychaetes (Wilson 1968; Williams 1964; Kirchman et al. 1982a), molluscs (Bayne 1969), barnacles (Knight-Jones 1953), bryozoans (Crisp and Williams 1960), and echinoderms (Highsmith 1982). Settlement can be stimulated by water-soluble neuroactive agents (D. E. Morse et al. 1979; Pawlik 1990) or soluble preparations of inductive substrata (Veitch and Hidu 1971; Müller 1973). Molluscan larvae have been stimulated to settle in enclosed volumes of seawater containing prey species (Thompson 1958) or conspecifics (Hidu 1969) without contacting the respective substrata. Given the turbulent advective processes of waves, tides, and currents, however, soluble compounds are unlikely to be present under natural conditions in sufficient concentrations to influence larvae, except at or very near the surface of the substratum (Crisp 1965; Denny and Shibata 1983). In fact, the few naturally occurring settlement cues that have been isolated and identified are insoluble in seawater (Kato et al. 1975; Yvin et al. 1985; Pawlik 1986).

Nevertheless, soluble compounds are thought to affect settlement in some species. Larvae of the mud snail *Ilyanassa obsoleta* (= *Nassarius obsoletus*) respond to a soluble factor emanating from sediments. In restricted embayments this factor may reach levels that effect cessation of swimming and onset of substratum exploration (Scheltema 1961). Larvae of the coral-eating nudibranch *Phestilla sibogae* may respond to a soluble coral-produced factor as they pass over shallow reefs (Hadfield and Scheuer 1985), although they may not respond until they are very near the substratum (Hadfield and Miller 1987). Similarly, larvae of another nudibranch, *Onchidoris bilamellata*, may begin substratum exploration after perceiving soluble cues associated with their barnacle prey (Chia and Koss 1988).

Marsden (1987) performed laboratory experiments with larvae of the tube worm *Spirobranchus giganteus*, an obligate associate of corals. She noted that in small experimental chambers, precompetent larvae swam toward some coral species in preference to other species, coral rubble, or control seawater, and she suggested that larvae may respond to a chemical cue diffusing from the preferred coral species (Marsden and Meeuwig 1990). In concert with responses to light, this preference would tend to entrain precompetent larvae near the reef, pending their maturation. It has not been demonstrated, however, that the coral diffusate has any effect on larvae at natural concentrations, or that larvae maintain their positions, let alone swim in a directed fashion, under natural conditions of water motion.

Chemotaxis toward a preferred substratum in flowing water has not been documented for the larvae of any marine invertebrate. Crisp (1965) made two persuasive arguments against larval chemotaxis. First, turbulent water flow over a substratum releasing a diffusing inducer would dilute the cue to negligible concentrations within a short distance from the surface. The factor would be present in perceptible quantities only in the viscous boundary layer adjacent to the surface. The depth of the boundary layer under natural flow conditions would be similar to the size of the larva; hence, larval response to a diffusing cue would essentially be contact chemoreception. Second, larvae are small enough to make orientation and navigation in a concentration gradient difficult. Larvae could detect a chemical gradient in one of two ways: (1) by perceiving a concentration difference between sensory organs placed some distance apart—for example, on opposite ends of the larval body; or (2) by integrating concentration changes as the larva moves through the water. The former strategy is unlikely because a concentration difference across the body length of a larva is likely to be imperceptible. The latter is equally implausible because poorly swimming larvae subjected to flow characterized by low Reynolds numbers are more apt to travel along with a mass of water than through it. The sperm of algae (Maier and Müller 1986) and some invertebrates (Miller and King 1983) are attracted to eggs by chemotaxis, but the process occurs over a distance measured in micrometers and within the viscous boundary layer surrounding the egg. Larvae of coral reef fishes appear to respond to diffusable chemical cues emanating from adult conspecifics and heterospecifics (Sweatman 1988), but fish larvae are larger and better swimmers than invertebrate larvae. Juvenile benthic invertebrates apparently respond to soluble cues with directed movements (Rittschof et al. 1983), and chemotaxis certainly occurs among adult marine invertebrates (Atema et al. 1988), which may obtain directional information from temporal patterns of diffusing chemical signals released into turbulent water flow (Moore and Atema 1988).

Chemosensory Organs

Sensory organs transduce environmental cues (light, gravity, or mechanical or chemical stimuli) into signals within the organism (neurotransmitters, hormones, or electrical impulses). The sensory organs of invertebrate larvae are poorly known (Laverack 1974; Chia and Rice 1978; Lacalli 1981; Burke 1983; Chia 1989). Their perceptive functions are usually presumptive, inferred from observations of larval behavior or based on histological and ultrastructural evidence. Their small size and delicate nature make larvae poor subjects for neurophysiological investigations. Moreover, because competent larvae are poised for metamorphosis, which usually results in drastic morphological changes, the sensory organs involved in settlement may fulfill their function only at the onset of this crucial transition and may not be subject to repeated experimentation. Considering these difficulties, it is not surprising that no larval chemoreceptive organs that respond to identified chemical signals have been unambiguously characterized, although some recent research shows promise toward this end (Chia and Koss 1988; Arkett et al. 1989).

It is important to note that stimulation of an epithelial sensory organ is not necessarily a requirement for the onset of settlement. There is good evidence that the larval nervous system can be directly influenced to trigger metamorphosis. This has been achieved by exposing larvae to seawater with an altered ionic composition (Baloun and Morse 1984; Yool et al. 1986; see above) and to various neuroactive agents (see above), and by direct electrical stimulation (Cameron and Hinegardner 1974; Satterlie and Cameron 1985). Compounds that "shock" larvae (e.g., H_2S) may cause settlement without the involvement of a specific chemoreceptor. There is more evidence, however, that epithelial chemosensory organs play a direct role in larval substratum selection under natural conditions.

The antennules of the cyprid larvae of barnacles were perhaps the first larval structures recognized to have a chemosensory role in selective settlement (Knight-Jones 1953). Cyprid larvae use brushlike disks attached to their antennules to walk over potential substrata at the time of settlement (Nott 1969). Once the settlement site has been chosen, the disks exude a permanent cement and metamorphosis follows. Nott and Foster (1969) described the structure of the attachment disks in some detail. In addition to a complex internal array of muscles, ducts, and glands, each disk bears a battery of sensory hairs that project beyond its brushlike surface. Three of these sensory hairs have exposed processes and are thought to function as chemoreceptors. Adjacent to the attachment disk, on the fourth segment of the antennule, are additional hairs that may play a chemosensory role (Gibson and Nott 1971). Nott and Foster (1969) proposed a mechanism by which the attachment disks might be employed in detecting the arthropodin cue of conspecific adults: the disks expel proteases as they contact the substratum, resulting in the localized breakdown of arthropodin and the release of characteristic amino acids that are then recognized by the sensory hairs. As I mentioned earlier, an alternative view, the "tactile chemical sense," was elaborated by Crisp and Meadows (1963; Crisp 1974, 1984), who proposed an analogy to an antigen-antibody reaction whereby the attachment disk adheres to the substratum-bound inducer by noncovalent bonding. In this scenario, chemoreception per se would not be involved. This idea has gained support through studies of the temporary attachment of cyprids to variously treated surfaces (Yule and Crisp 1983) and with the discovery of sticky proteins exuded by the attachment disk during reversible attachment (Walker and Yule 1984; Yule and Walker 1987).

The presumptive chemosensory organs of molluscan larvae have been described for a few species. Bonar (1978) characterized the cephalic sensory organ of veliger larvae of the obligate coral-eating nudibranch *Phestilla sibogae*. The organ, located between the lobes of the velum, is made up of three types of cells. Of these, flask-shaped ciliated cells with direct axonal connections to the larval nervous system may have a chemosensory function. Chia and Koss (1982) described the putative sensory organs of the larvae of *Rostanga pulchra*, a nudibranch that settles specifically on its sponge prey, *Ophlitaspongia pennata*. Competent veligers of *R. pulchra* bear two cylindrical rhinophores between their velar lobes. The core of each rhinophore contains a ganglion from which dendritic terminals form tufts at the apex. These dendritic endings were postulated to serve a chemosensory function. The veligers of a third nudibranch, the barnacle predator *Onchidoris bilamellata*, possess two sensory fields on the surface of the larval propodium (the structure used for crawling, which later becomes part of the adult foot; Chia and Koss 1989). Sensory cells of each field are directly innervated by a pair of ganglia located just below the propodial epidermis. The propodial ganglia and their associated sensory structures were hypothesized to detect the chemical factor produced by adult barnacles that stimulates cessation of swimming and onset of substratum exploration (Chia and Koss 1988; Chia 1989). Arkett et al. (1989) have made preliminary intracellular recordings from cells in the sensory field of *O. bilamellata* larvae. These cells respond to barnacle-conditioned seawater with a slow, small-amplitude depolarization but do not respond to control seawater. Injection with lucifer yellow revealed the depolarizing cells to be flask shaped with dendritic processes extending to the propodial surface.

Based on the responses of larvae to various pharmacological and neuroactive agents, Morse and colleagues have proposed a complex dual-pathway system controlling the induction of settlement and metamorphosis of the abalone *Haliotis rufescens* (D. E. Morse 1990; see above). The receptors for both the natural settlement inducer (derived from encrusting red algae) and the waterborne amino acids that regulate this induction may be located on epithelial chemosensory cilia. D. E. Morse et al. (1980a) described the secretion of glycopeptides from the cephalic sensory organ as larvae underwent "behavioral metamorphosis" in response to GABA, a neurotransmitter thought to act at the same receptor as the naturally occurring inducer. After exuding glycopeptides, larvae subsequently shed the ciliated cells of the velum. Trapido-Rosenthal and Morse (1986) determined that a radiolabeled settlement inducer, baclofen (an analogue of GABA), became dissociated from metamorphosing larvae after 20 hours of incubation, evidently coinciding with the loss of velar cells and other epithelial cilia. The authors concluded that chemosensory receptors specific for the induction of settlement were present on these cilia. Purification of mRNA from preparations of cilia followed by cDNA synthesis and amplification revealed mRNA sequences that apparently code for specific proteins involved in transducing the chemosensory signal (Wodicka and Morse 1991). It is not clear, however, whether these cilia are from the velum, the cephalic sensory organ, or elsewhere on the larval surface.

Barlow (1990) noted that GABA effected a cessation of velar ciliary beat in both precompetent and competent larvae of *H. rufescens*. Restrained larvae exposed to GABA exhibited foot movements that were thought to be a component of normal settlement behavior. Using intracellular recording techniques, Barlow monitored electrical responses of velar cells to GABA and determined that the receptor for the compound was most likely not present on velar cells, though her results did not necessarily preclude their presence on velar cilia. She pointed out that the location of putative GABA receptors is unclear: they may be internal (synaptic) or epithelial (chemoreceptive).

Among polychaetes, the larval sensory organs involved in settlement are perhaps best characterized for the reef-building sabellariids of the genus *Phragmatopoma*. Eckelbarger (1978) described tufts of cilia scattered over the bodies of *Phragmatopoma lapidosa lapidosa* larvae and suggested that these structures may have a role in substratum selection because they are concentrated on parts of the body used when exploring the substratum. Amieva and Reed (1987; Amieva et al. 1987) used video microscopy to study the behavior of *P. l. californica* at settlement and examined the ultrastructure of the larval tentacles of this subspecies. They determined that the tufts of cilia on the tentacle surfaces are immotile and borne by cells in direct communication with the larval nervous system. The morphological evidence and observations of settlement behavior led them to suggest that the ciliary tufts may be involved in the perception of settlement cues.

Ultrastructural studies of marine bryozoan larvae have revealed several structures that may have sensory functions (Reed 1988a; Reed et al. 1988). The putative larval chemosensory organ consists of a long bundle of cilia, called the vibratile plume, which projects from the larva's anterior midline (Reed 1988b). The ciliary bundle is attached to a glandular and sensory complex, the pyriform organ. The cells bearing the vibratile plume appear to be directly innervated. The chemosensory role of the vibratile plume and pyriform organ is corroborated by observations of larval behavior at settlement: bryozoan larvae press their pyriform organs against potential substrata and beat the surface with their vibratile plumes.

Less is known of potential larval chemosensory organs in other invertebrate phyla. Vandermeulen (1974) and Chia and Koss (1979) described putative sensory cells that might mediate substratum selection in the planulae of a coral and an anemone, respectively. Burke (1980) reported sensory cells on the tube feet of the adult rudiment of competent echinoid larvae that may detect cues as the tube feet contact the substratum. Ascidian tadpole larvae bear peripheral ciliated sensory neurons in the anterior adhesive papillae that may detect chemical properties of the substratum (Torrence and Cloney 1988).

Comparisons with Terrestrial Insects

The chemical ecology of terrestrial insects is much better understood than that of marine invertebrates. Compounds responsible for various insect behaviors have been isolated and identified, the chemosensory structures involved in signal perception have been well characterized through the use of electrophysiological techniques, and the neuroendocrine mechanisms controlling development have been intensively studied (Hansen 1978; Vinnikov 1982; Downer and Laufer 1983; Bell and Cardé 1984; Prestwich 1985; O'Connell 1986).

With regard to larval ecology and substratum selection, there are pronounced differences between insects and benthic marine invertebrates. Among most terrestrial insects the adult is the dispersive phase in the life history, and the larva is sedentary. The diets of adult and larval insects may differ radically, much like their marine counterparts, but the food selected for larval growth is chosen by the egg-laying adult: substratum selection is an adult, rather than a larval, concern (Fig. 6.1). Gregariousness, with all its concomitant advantages, occurs commonly among insects, but for most species it is restricted to adults. Substratum selection, aggregation, and mate location are all mediated by chemical signals in insects, and the cues may be volatile or surface associated, perceived by olfactory or tactile chemoreceptors, respectively. Finally, unlike marine invertebrates, chemical signals involved in the onset of insect metamorphosis are largely endogenous (hormones) rather than exogenous.

Substratum selection by ovipositing adult insects is common. Mobile females are generally stimulated to lay eggs only in response to specific host-produced natural products. For example, pierid butterflies oviposit only on plants of the family Cruciferae that produce glucosinolates (Chew 1977), sawflies lay eggs only on willows that synthesize specific phenolic glycosides (Roininen and Tahvanainen 1989),

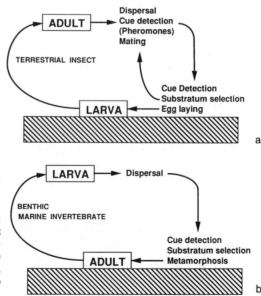

Figure 6.1 Comparison of the generalized life histories of (a) a terrestrial insect and (b) a benthic marine invertebrate. Terrestrial insect larvae are sedentary, and dispersal and substratum selection are undertaken by the adults; among marine invertebrates, however, it is the larvae that disperse and select the final substrate, where they metamorphose into sendentary adults.

and some parasitic wasps are stimulated to oviposit on caterpillars by specific hydrocarbons produced by the host (Vinson 1984). Larvae of these species are essentially sessile; they are unlikely to find another host if the one chosen by the ovipositing female is unsuitable (Chew 1977). Aggregation pheromones similarly act on adult, rather than larval, insects, and their production is often a response to the presence of food (e.g., beetles: Birgersson et al. 1988; Oehlschlager et al. 1988).

Insect olfactory cues have received particular attention (Kramer 1978; O'Connell 1986). Volatile compounds are used as sex attractants; as trail markers; as alarm, aggregation, or repulsion pheromones; and as indicators of food availability. Chemical communication of this kind may be very complex, requiring suites of several individual compounds in specific ratios. Contact chemoreception is also common among insects, especially in food and mate recognition (Städler 1984; Prestwich 1985). The sensory structures (sensilla) responsible for both olfaction and contact chemoreception have been characterized morphologically and electrophysiologically (Hansen 1978; Städler 1984).

The sensory organs associated with the antennules of barnacle cyprid larvae have been compared to those of other arthropods, including insects (Nott and Foster 1969; Gibson and Nott 1971). Some authors have compared chemically mediated substratum selection of invertebrate larvae to pheromonal communication by terrestrial insects (Burke 1984, 1986; Pawlik and Faulkner 1986, 1988), although insects differ in being able to detect volatile pheromones at great distances from their sources. Gregarious settlement of the polychaetes *Phragmatopoma lapidosa californica* and *P. l. lapidosa* was induced on larval contact with specific free fatty acids isolated from the tubes of adult worms (Pawlik and Faulkner 1986; Pawlik 1988b); among insects, specific fatty acids stimulate electrophysiological responses in sensory receptors of *Necrophorous* beetles (Boeckh 1962) and fleshflies (Shimada 1978).

Metamorphosis is under the control of endogenous chemical signals in both hemi- and holometabolous insects, although environmental cues such as temperature and photoperiod play a role in timing the event. A balance of competing hormones—ecdysone and juvenile hormone—produced by the prothoracic gland and brain, respectively, dictate the advancement of the larva through successive molts, pupation, and to adulthood (Downer and Laufer 1983). In barnacles, synthetic analogues of insect juvenile hormones trigger metamorphosis without attachment (Gomez et al. 1973; Freeman and Costlow 1983), and compounds similar to insect juvenile hormones have been identified in extracts of crustaceans (Laufer et al. 1987). Fouling barnacles could potentially be controlled by disrupting normal development through the manipulation of the endocrine system (Fingerman 1988), a method already used to control pest insects.

Advances in the study of insect chemical ecology have led to the use of pheromone analogues in pest control (O'Connell 1986). Similar potential may exist for research into the chemical ecology of marine invertebrates. For example, parasitic rhizocephalan barnacles attack commercially important crabs worldwide and prevent them from reproducing or reaching marketable size; their impact on crustacean fisheries can be considerable (Lester 1978; P. T. Johnson et al. 1986). Isolation and identification of the compounds responsible for the localization of host crabs by female cyprids, or virginal externae by male cyprids, could lead to control measures similar to those used to manage some insect pests.

Acknowledgments   Earlier versions of this review were commented on by L. Barlow, C. A. Butman, G. Gibson, J. P. Grassle, H. Hess, R. Koss, C. G. Reed, and M. Strathmann. Special thanks go to D. Manker and T. F. Molinski for sending important references. Support for this undertaking was provided by a Killam Memorial Postdoctoral Fellowship through the University of Alberta, Edmonton, Canada, by a Woods Hole Oceanographic Institution Postdoctoral Fellowship, and by a National Science Foundation Presidential Young Investigator Award through the University of North Carolina at Wilmington.

References

Agius, L. 1979. Larval settlement in the echiuran worm *Bonellia viridis*: settlement on both the adult proboscis and body trunk. Mar. Biol. 53:125–129.

Agius, L., Ballantine, J.A., Ferrito, V., Jaccarini, V., Murray-Rust, P., Pelter, A., Psaila, A.F., and Schembri, P.J. 1979. The structure and physiological activity of bonellin—a unique chlorin derived from *Bonellia viridis*. Pure Appl. Chem. 51:1847–1864.

Amieva, M.R., and Reed, C.G. 1987. Functional morphology of the larval tentacles of *Phragmatopoma californica* (Polychaeta: Sabellariidae): composite larval and adult organs of multifunctional significance. Mar. Biol. 95:243–258.

Amieva, M.R., Reed, C.G., and Pawlik, J.R. 1987. Ultrastructure and behavior of the larva of *Phragmatopoma californica* (Polychaeta: Sabellariidae): identification of sensory organs potentially involved in substrate selection. Mar. Biol. 95:259–266.

Arkett, S.A., Chia, F.S., Goldberg, J.I., and Koss, R. 1989. Identified settlement receptor cells in a nudibranch veliger respond to specific cue. Biol. Bull. 176:155–160.

Atema, J., Fay, R.R., Popper, A.N., and Tavolga, W.N., eds. 1988. Sensory biology of aquatic animals. New York: Springer Verlag.

Bahamondes-Rojas, I., and Dherbomez, M. 1990. Purification partielle des substances glycoconjuguées capables d'induire la métamorphose des larves compétentes d'*Eubranchus doriae* (Trinchèse, 1879), mollusque nudibranche. J. Exp. Mar. Biol. Ecol. 144:17–27.

Ballantine, J.A., Psaila, A.F., Pelter, A., Murray-Rust, P., Ferrito, V., Schembri, P., and Jaccarini, V. 1980. The structure of bonellin and its derivatives. Unique physiologically active chlorins from the marine echiuran *Bonellia viridis*. J. Chem. Soc. Perkin Trans. I, pp. 1080–1089.

Baloun, A.J., and Morse, D.E. 1984. Ionic control of settlement and metamorphosis in larval *Haliotis rufescens* (Gastropoda). Biol. Bull. 167:124–138.

Barker, M.F. 1977. Observations on the settlement of the brachiolaria larvae of *Stichaster australis* (Verrill) and *Coscinasterias calamaria* (Gray)(Echinodermata: Asteroidea) in the laboratory and on the shore. J. Exp. Mar. Biol. Ecol. 30:95–108.

Barlow, L.A. 1990. Electrophysiological and behavioral responses of larvae of the red abalone (*Haliotis rufescens*) to settlement-inducing substances. Bull. Mar. Sci. 46:537–554.

Barnes, J.R., and Gonor, J.J. 1973. The larval settling response of the lined chiton *Tonicella lineata*. Mar. Biol. 20:259–264.

Baxter, G., and Morse, D.E. 1987. G protein and diacylglycerol regulate metamorphosis of planktonic molluscan larvae. Proc. Natl. Acad. Sci. USA 84:1867–1870.

Bayne, B.L. 1969. The gregarious behaviour of the larvae of *Ostrea edulis* L. at settlement. J. Mar. Biol. Assoc. U.K. 49:327–356.

Bell, W.J., and Cardé, R.T., eds. 1984. Chemical ecology of insects. Sunderland, Mass.: Sinauer Associates.

Birgersson, G., Schlyter, F., Bergström, G., and Löfqvist, J. 1988. Individual variation in aggregation pheromone content of the bark beetle, *Ips typographus*. J. Chem. Ecol. 14:1737–1761.

Boeckh, J. 1962. Elektrophysiologische Untersuchungen an einzelnen Geruchsrezeptoren auf den Antennen des Totengräbers (*Necrophorus*, Coleoptera). Z. Vgl. Physiol. 46:212–248.

Bonar, D.B. 1978. Ultrastructure of a cephalic sensory organ in larvae of the gastropod *Phestilla sibogae* (Aeolidacea, Nudibranchia). Tissue & Cell 10:153–165.

Bonar, D.B., Coon, S.L., Walch, M., Weiner, R.M., and Fitt, W. 1990. Control of oyster settlement and metamorphosis by endogenous and exogenous chemical cues. Bull. Mar. Sci. 46:484–498.

Bonar, D.B., Coon, S.L., Weiner, R.M., and Colwell, R.R. 1985. Induction of oyster metamorphosis by bacterial products and biogenic amines. Bull. Mar. Sci. 37:763.

Burke, R.D. 1980. Podial sensory receptors and the induction of metamorphosis in echinoids. J. Exp. Mar. Biol. Ecol. 47:223–234.

Burke, R.D. 1983. The induction of metamorphosis of marine invertebrate larvae: stimulus and response. Can. J. Zool. 61:1701–1719.

Burke, R.D. 1984. Pheromonal control of metamorphosis in the Pacific sand dollar, *Dendraster excentricus*. Science 225:442–443.

Burke, R.D. 1986. Pheromones and the gregarious settlement of marine invertebrate larvae. Bull. Mar. Sci. 39:323–331.

Bushek, D. 1988. Settlement as a major determinant of intertidal oyster and barnacle distributions along a horizontal gradient. J. Exp. Mar. Biol. Ecol. 122:1–18.

Butman, C.A. 1987. Larval settlement of soft-sediment invertebrates: the spatial scales of pattern explained by active habitat selection and the emerging role of hydrodynamical processes. Oceanogr. Mar. Biol. Annu. Rev. 25:113–165

Butman, C.A., Grassle, J.P., and Webb, C.M. 1988. Substrate choices made by marine larvae settling in still water and in a flume flow. Nature (Lond.) 333:771–773.

Cameron, R.A., and Hinegardner, R.T. 1974. Initiation of metamorphosis in laboratory cultured sea urchins. Biol. Bull. 146:335–342.

Cariello, L., De Nicola Giudici, M., Zanetti, L., and Prota, G. 1978. Neobonellin, a new biologically active pigment from *Bonellia viridis*. Experientia 34:1427–1429.

Chabot, R., and Bourget, E. 1988. Influence of substratum heterogeneity and settled barnacle density on the settlement of cypris larvae. Mar. Biol. 97:45–56.

Chew, F.S. 1977. Coevolution of pierid butterflies and their cruciferous foodplants. II. The distribution of eggs on potential foodplants. Evolution 31:568–579.

Chia, F.S. 1989. Differential larval settlement of benthic marine invertebrates. *In* Reproduction, genetics and distributions of marine organisms, ed. J.S. Ryland and P.A. Tyler, pp. 3–12. Fredensborg, Denmark: Olsen and Olsen.

Chia, F.S., and Koss, R. 1979. Fine structural studies of the nervous system and the apical organ in the planula larva of the sea anemone *Anthopleura elegantissima*. J. Morphol. 160:275–297.

Chia, F.S., and Koss, R. 1982. Fine structure of the larval rhinophores of the nudibranch, *Rostanga pulchra*, with emphasis on the sensory receptor cells. Cell Tissue Res. 225:235–248.

Chia, F.S., and Koss, R. 1988. Induction of settlement and metamorphosis of the veliger larvae of the nudibranch *Onchidoris bilamellata*. Int. J. Invertebr. Reprod. Dev. 14:53–70.

Chia, F.S., and Koss, R. 1989. The fine structure of the newly discovered propodial ganglia of the veliger larva of the nudibranch *Onchidoris bilamellata*. Cell Tissue Res. 256:17–26.

Chia, F.S., and Rice, M.E., eds. 1978. Settlement and metamorphosis of marine invertebrate larvae. New York: Elsevier.

Coale, K.H., Chin, C.S., Massoth, G.J., Johnson, K.S., and Baker, E.T. 1991. In situ chemical mapping of dissolved iron and manganese in hydrothermal plumes. Nature (Lond.) 352:325–328.

Cochard, J.C., Chevolot, L., Yvin, J.C., and Chevolot-Magueur, A.M. 1989. Induction de la metamorphose de la coquille Saint Jacques *Pecten maximus* L. par des derives de la tyrosine extraits de l'algue *Delesseria sanguinea* Lamouroux ou synthetiques. Haliotis 19:129–154.

Colman, J.S. 1933. The nature of the intertidal zonation of plants and animals. J. Mar. Biol. Assoc. U.K. 18:435–476.

Connell, J.H. 1985. The consequences of variation in initial settlement vs. post-settlement mortality in rocky intertidal communities. J. Exp. Mar. Biol. Ecol. 93:11–45.

Coon, S.L., and Bonar, D.B. 1987. The role of DOPA and dopamine in oyster settlement behavior. Am. Zool. 27:128A.

Coon, S.L., Bonar, D.B., and Weiner, R.M. 1985. Induction of settlement and metamorphosis of the Pacific oyster, *Crassostrea gigas* (Thunberg), by L-DOPA and catecholamines. J. Exp. Mar. Biol. Ecol. 94:211–221.

Coon, S.L., Fitt, W.K., and Bonar, D.B. 1990a. Competence and delay of metamorphosis in the Pacific oyster *Crassostrea gigas*. Mar. Biol. 106:379–387.

Coon, S.L., Walch, M., Fitt, W.K., Bonar, D.B., and Weiner, R.M. 1988. Induction of settlement behavior in oyster larvae by ammonia. Am. Zool. 28:70A.

Coon, S.L., Walch, M., Fitt, W.K., Weiner, R.M., and Bonar, D.B. 1990b. Ammonia induces settlement behavior in oyster larvae. Biol. Bull. 179:297–303.

Cooper, K. 1982. A model to explain the induction of settlement and metamorphosis of planktonic eyed-pediveligers of the blue mussel *Mytilus edulis* L. by chemical and tactile cues. J. Shellfish Res. 2:117.

Crisp, D.J. 1965. Surface chemistry, a factor in the settlement of marine invertebrate larvae. Bot. Goth. 3:51–65.

Crisp, D.J. 1967. Chemical factors inducing settlement in *Crassostrea virginica* (Gmelin). J. Anim. Ecol. 36:329–335.

Crisp, D.J. 1974. Factors influencing the settlement of marine invertebrate larvae. *In* Chemoreception in marine organisms, ed. P.T. Grant and A.M. Mackie, pp. 177–265. New York: Academic Press.

Crisp, D.J. 1979. Dispersal and re-aggregation in sessile marine invertebrates, particularly barnacles. *In* Marine organisms—genetics, ecology and evolution, vol. 11, ed. G. Larwood and B.R. Rosen, pp. 319–327. London: Academic Press.

Crisp, D.J. 1984. Overview of research on marine invertebrate larvae, 1940–1980. *In* Marine biodeterioration: an interdisciplinary study, ed. J.D. Costlow and R.C. Tipper, pp. 103–126. Annapolis, Md.: Naval Institute Press.

Crisp, D.J. 1990. Gregariousness and systematic affinity in some North Carolinian barnacles. Bull. Mar. Sci. 47:516–525.

Crisp, D.J., and Meadows, P.S. 1962. The chemical basis of gregariousness in cirripedes. Proc. R. Soc. Lond. B 156:500–520.

Crisp, D.J., and Meadows, P.S. 1963. Adsorbed layers: the stimulus to settlement in barnacles. Proc. R. Soc. Lond. B 158:364–387.

Crisp, D.J., and Williams, G.B. 1960. Effect of extracts from fucoids in promoting settlement of epiphytic Polyzoa. Nature (Lond.) 188:1206–1207.

Culliney, J.L. 1972. Settling of larval shipworms, *Teredo navalis* L. and *Bankia gouldi* Bartsch, stimulated by humic material (Gelbstoff). *In* Proc. Third Int. Congr. Mar. Corrosion and Fouling, pp. 622–629. Evanston, Ill.: Northwestern University Press.

Cuomo, M.C. 1985. Sulphide as a larval settlement cue for *Capitella* sp. I. Biogeochemistry 1:169–181.

De Nicola Giudici, M. 1984. Defence mechanism of *Bonellia viridis*. Mar. Biol. 78:271–273.

Denny, M.W., and Shibata, M.F. 1989. Consequences of surf-zone turbulence for settlement and external fertilization. Am. Nat. 134:859–889.

Dirnberger, J.M. 1990. Benthic determinants of settlement for planktonic larvae: availability of settlement sites for the tube-building polychaete *Spirorbis spirillum* (Linnaeus) settling onto seagrass blades. J. Exp. Mar. Biol. Ecol. 140:89–105.

Donaldson, S. 1974. Larval settlement of a symbiotic hydroid: specificity and nematocyst responses in planulae of *Proboscidactyla flavicirrata*. Biol. Bull. 147:573–585.

Downer, R.G.H., and Laufer, H., eds. 1983. Endocrinology of insects. New York: Alan R. Liss.

Dubilier, N. 1988. H_2S—a settlement cue or a toxic substance for *Capitella* sp. I larvae? Biol. Bull. 174:30–38.

Eckelbarger, K.J. 1978. Metamorphosis and settlement in the Sabellariidae. *In* Settlement and metamorphosis of marine invertebrate larvae, ed. F.S. Chia and M.E. Rice, pp. 145–164. New York: Elsevier.

Eyster, L.S., and Pechenik, J.A. 1987. Attachment of *Mytilus edulis* L. larvae on algal and byssal filaments is enhanced by water agitation. J. Exp. Mar. Biol. Ecol. 114:99–110.

Faulkner, D.J., and Ghiselin, M.T. 1983. Chemical defense and evolutionary ecology of dorid nudibranchs and some other opisthobranch gastropods. Mar. Ecol. Prog. Ser. 13:295–301.

Fingerman, M. 1988. Application of endocrine manipulations to the control of marine fouling crustaceans. *In* Marine biodeterioration. Advanced techniques applicable to the Indian Ocean, ed. M.F. Thompson, R. Sarojini, and R. Nagabhushanam, pp. 81–91. New Delhi: Oxford and IBH Publishing.

Fitt, W.K., Coon, S.L., Walch, M., Weiner, R.M., Colwell, R.R., and Bonar, D.B. 1990. Settlement behavior and metamorphosis of oyster larvae (*Crassostrea gigas*) in response to bacterial supernatants. Mar. Biol. 106:389–394.

Fitt, W.K., and Hoffman, D.K. 1985. Chemical induction of settlement and metamorphosis of planulae and buds of the reef-dwelling coelenterate *Cassiopeia andromeda*. *In* Proc. Fifth Int. Coral Reef Symp., vol. 5, pp. 239–244.

Fitt, W.K., Hoffmann, D.K., Wolk, M., and Rahat, M. 1987. Requirement of exogenous inducers for metamorphosis of axenic larvae and buds of *Cassiopeia andromeda* (Cnidaria: Scyphozoa). Mar. Biol. 94:415–422.

Freeman, J.A., and Costlow, J.D. 1983. The cyprid molt cycle and its hormonal control in the barnacle *Balanus amphitrite*. J. Crustacean Biol. 3:173–182.

Gabbott, P.A., and Larman, V.N. 1987. The chemical basis of gregariousness in cirripedes: a review (1953–1984). *In* Barnacle biology, ed. A.J. Southward, pp. 377–388. Rotterdam: A.A. Balkema.

Gaines, S., and Roughgarden, J. 1985. Larval settlement rate: a leading determinant of structure in an ecological community of the marine intertidal zone. Proc. Natl. Acad. Sci. USA 82:3707–3711.

Gee, J.M. 1964. Chemical stimulation of settlement in larvae of *Spirorbis rupestris* (Serpulidae). Anim. Behav. 13:181–186.

Gibson, P.H., and Nott, J.A. 1971. Concerning the fourth antennular segment of the cypris larva of *Balanus balanoides*. *In* Proc. Fourth Eur. Mar. Biol. Symp., ed. D.J. Crisp, pp. 227–236. Cambridge: Cambridge University Press.

Gomez, E.D., Faulkner, D.J., Newman, W.A., and Ireland, C. 1973. Juvenile hormone mimics: effect on cirriped crustacean metamorphosis. Science 179:813–814.

Grassle, J.P., and Butman, C.A. 1989. Active habitat selection by larvae of the polychaetes, *Capitella* spp. I and II, in a laboratory flume. *In* Reproduction, genetics and distributions of marine organisms, ed. J.S. Ryland and P.A. Tyler, pp. 107–114. Fredensborg, Denmark: Olsen and Olsen.

Gray, J.S. 1974. Animal-sediment relationships. Oceanogr. Mar. Biol. Annu. Rev. 12:223–261.

Hadfield, M.G. 1976. Molluscs associated with living tropical corals. Micronesica 12:133–148.

Hadfield, M.G. 1977. Chemical interactions in larval settling of a marine gastropod. *In* Marine natural products chemistry, ed. D.J. Faulkner and W.H. Fenical, pp. 403–413. New York: Plenum Press.

Hadfield, M.G. 1978. Metamorphosis in marine molluscan larvae: an analysis of stimulus and response. *In* Settlement and metamorphosis of marine invertebrate larvae, ed. F.S. Chia and M.E. Rice, pp. 165–175. New York: Elsevier.

Hadfield, M.G. 1984. Settlement requirements of molluscan larvae: new data on chemical and genetic roles. Aquaculture 39:283–298.

Hadfield, M.G., and Miller, S.E. 1987. On developmental patterns of opisthobranchs. Am. Malacol. Bull. 5:197–214.

Hadfield, M.G., and Pennington, J.T. 1990. The nature of the metamorphic signal and its internal transduction in larvae of the nudibranch *Phestilla sibogae*. Bull. Mar. Sci. 46:455–464.

Hadfield, M.G., and Scheuer, D. 1985. Evidence for a soluble metamorphic inducer in *Phestilla*: ecological, chemical and biological data. Bull. Mar. Sci. 37:556–566.

Hahn, K.O. 1989. Induction of settlement in competent abalone larvae. *In* Handbook of culture of abalone and other marine gastropods, ed. K.O. Hahn, pp. 101–112. Boca Raton, Fla.: CRC Press.

Hansen, K. 1978. Insect chemoreception. *In* Receptors and recognition, series B, vol. 5, of Taxis and behavior, elementary sensory systems in biology, ed. G.L. Hazelbauer, pp. 231–292. London: Chapman and Hall.

Harington, C.R. 1921. A note on the physiology of the ship-worm (*Teredo norvegica*). Biochem. J. 15:736–741.

Havenhand, J.N., and Svane, I. 1989. Larval behaviour, recruitment, and the rôle of adult attraction in *Ascidia mentula* O.F. Müller. *In* Reproduction, genetics and distributions of marine organisms, ed. J.S. Ryland and P.A. Tyler, pp. 127–132. Fredensborg, Denmark: Olsen and Olsen.

Hidu, J. 1969. Gregarious setting in the American oyster *Crassostrea virginica* Gmelin. Chesapeake Sci. 10:85–92.

Highsmith, R.C. 1982. Induced settlement and metamorphosis of sand dollar (*Dendraster excentricus*) larvae in predator-free sites: adult sand dollar beds. Ecology 63:329–337.

Hill, E.M., and Holland, D.L. 1985. Influence of oil shale on intertidal organisms: isolation and characterization of metalloporphyrins that induce the settlement of *Balanus balanoides* and *Elminius modestus*. Proc. R. Soc. Lond. B 225:107–120.

Hirata, K.Y., and Hadfield, M.G. 1986. The role of choline in metamorphic induction of *Phestilla* (Gastropoda, Nudibranchia). Comp. Biochem. Physiol. C 84:15–21.

Høeg, J., and Lützen, J. 1985. Crustacea. Rhizocephala. *In* Marine invertebrates of Scandinavia, no. 6. Olso: Norwegian University Press.

Holland, D.L., Crisp, D.J., Huxley, R., and Sisson, J. 1984. Influence of oil shale on intertidal organisms: effect of oil shale extract on settlement of the barnacle *Balanus balanoides* (L.). J. Exp. Mar. Biol. Ecol. 75:245–255.

Huxley, R., Holland, D.L., Crisp, D.J., and Smith, R.S.L. 1984. Influence of oil shale on intertidal organisms: effect of oil shale surface roughness on settlement of the barnacle *Balanus balanoides* (L.). J. Exp. Mar. Biol. Ecol. 82:231–237.

Jaccarini, V., Agius, L., Schembri, P.J., and Rizzo, M. 1983. Sex determination and larval sexual interaction in *Bonellia viridis* Rolando (Echiura: Bonelliidae). J. Exp. Mar. Biol. Ecol. 66:25–40.

Jaeckle, W.B., and Manahan, D.T. 1989. Feeding by a "nonfeeding" larva: uptake of dissolved amino acids from seawater by lecithotrophic larvae of the gastropod *Haliotis* rufescens. Mar. Biol. 103:87–94.

Jensen, R.A., and Morse, D.E. 1984. Intraspecific facilitation of larval recruitment: gregarious settlement of the polychaete *Phragmatopoma californica* (Fewkes). J. Exp. Mar. Biol. Ecol. 83:107–126.

Jensen, R.A., and Morse, D.E. 1990. Chemically induced metamorphosis of polychaete larvae in both the laboratory and ocean environment. J. Chem. Ecol. 16:911–930.

Jensen, R.A., Morse, D.E., Petty, R.L., and Hooker, N. 1990. Artificial induction of larval metamorphosis by free fatty acids. Mar. Ecol. Prog. Ser. 67:55–71.

Johnson, L.E., and Strathmann, R.R. 1989. Settling barnacle larvae avoid substrata previously occupied by a mobile predator. J. Exp. Mar. Biol. Ecol. 128:87–103.

Johnson, P.T., MacIntosh, R.A., and Somerton, D.A. 1986. Rhizocephalan infection in blue king crabs, *Paralithodes platypus*, from Olga Bay, Kodiak Island, Alaska. Fish. Bull. U.S. 84:177–184.

Jones, L.L. 1978. The life history patterns and host selection behavior of a sponge symbiont, *Membranobalanus orcutti* (Pilsbry) (Cirripedia). Ph.D. dissertation, University of California, San Diego.

Kato, T., Kumanireng, A.S., Ichinose, I., Kitahara, Y., Kakinuma, Y., Nishihira, M., and Kato, M. 1975. Active components of *Sargassum tortile* effecting the settlement of swimming larvae of *Coryne uchidai*. Experientia 31:433–434.

Keck, R., Maurer, D., Kauer, J.C., and Sheppard, W.A. 1971. Chemical stimulants affecting larval settlement in the American oyster. Proc. Natl. Shellfish. Assoc. 61:24–28.

Keough, M.J. 1983. Patterns of recruitment of sessile invertebrates in two subtidal habitats. J. Exp. Mar. Biol. Ecol. 66:213–245.

Kirchman, D., Graham, S., Reish, D., and Mitchell, R. 1982a. Bacteria induce settlement and metamorphosis of *Janua (Dexiospira) brasiliensis* Grube (Polychaeta: Spirorbidae). J. Exp. Mar. Biol. Ecol. 56:153–163.

Kirchman, D., Graham, S., Reish, D., and Mitchell, R. 1982b. Lectins may mediate the settlement and metamorphosis of *Janua (Dexiospira) brasiliensis* Grube (Polychaeta: Spirorbidae). Mar. Biol. Lett. 3:131–142.

Kiseleva, G.A. 1966. Factors stimulating larval metamorphosis of a lamellibranch, *Brachyodontes lineatus* (Gmelin). Zool. Zh. 45:1571–1572.

Knight-Jones, E.W. 1953. Laboratory experiments on gregariousness during setting in *Balanus balanoides* and other barnacles. J. Exp. Biol. 30:584–599.

Knight-Jones, E.W., and Stevenson, J.P. 1950. Gregariousness during settlement in the barnacle *Elminius modestus* Darwin. J. Mar. Biol. Assoc. U.K. 29:281–297.

Korringa, P. 1940. Experiments and observations on swarming, pelagic life and setting in the European flat oyster, *Ostrea edulis* L. Arch. Neerl. Zool. 5:1–249.

Kramer, E. 1978. Insect pheromones. *In* Receptors and recognition, series B, vol. 5, of Taxis and behavior, elementary sensory systems in biology, ed. G.L. Hazelbauer, pp. 205–229. London: Chapman and Hall.

Lacalli, T.C. 1981. Structure and development of the apical organ in trochophores of *Spirobranchus polycerus*, *Phyllodoce maculata* and *Phyllodoce mucosa* (Polychaeta). Proc. R. Soc. Lond. B 212:381–402.

Larman, V.N. 1984. Protein extracts from some marine animals which promote barnacle settlement: possible relationship between a protein component of arthropod cuticle and actin. Comp. Biochem. Physiol. B 77:73–81.

Larman, V.N., and Gabbott, P.A. 1975. Settlement of cyprid larvae of *Balanus balanoides* and *Elminius modestus* induced by extracts of adult barnacles and other marine animals. J. Mar. Biol. Assoc. U.K. 55:183–190.

Larman, V.N., Gabbott, P.A., and East, J. 1982. Physico-chemical properties of the settlement factor proteins from the barnacle *Balanus balanoides*. Comp. Biochem. Physiol. B 72:329–338.

Laufer, H., Borst, D., Baker, F.C., Carrasco, C., Sinkus, M., Reuter, C.C., Tsai, L.W., and Schooley, D.A. 1987. Identification of a juvenile hormone–like compound in a crustacean. Science 235:202–205.

Laverack, M.S. 1974. The structure and function of chemoreceptor cells. *In* Chemoreception in marine organisms, ed. P.T. Grant and A.M. Mackie, pp. 1–48. New York: Academic Press.

Lester, R.J.G. 1978. Marine parasites costly for fishermen. Aust. Fish. 37:32–33.

LeTourneux, F., and Bourget, E. 1988. Importance of physical and biological settlement cues used at different spatial scales by the larvae of *Semibalanus balanoides*. Mar. Biol. 97:57–66.

Levantine, P.L., and Bonar, D.B. 1986. Metamorphosis of *Ilyanassa obsoleta*: natural and artificial inducers. Am. Zool. 26(4):14A.

Lewis, C.A. 1978. A review of substratum selection in free-living and symbiotic cirripeds. *In* Settlement and metamorphosis of marine invertebrate larvae, ed. F.S. Chia and M.E. Rice, pp. 207–218. New York: Elsevier.

Lutz, R.A., Jablonski, D., and Turner, R.D. 1984. Larval development and dispersal at deep-sea hydrothermal vents. Science 226:1451–1454.

Lynch, W.F. 1961. Extrinsic factors influencing metamorphosis in bryozoan and ascidian larvae. Am. Zool. 1:59–66.

McGee, B.L., and Targett, N.M. 1989. Larval habitat selection in *Crepidula* (L.) and its effect on adult distribution patterns. J. Exp. Mar. Biol. Ecol. 131:195–214.

McGrath, D., King, P.A., and Gosling, E.M. 1988. Evidence for the direct settlement of *Mytilus edulis* larvae on adult mussel beds. Mar. Ecol. Prog. Ser. 47:103–106.

Maier, I., and Müller, D.G. 1986. Sexual pheromones in algae. Biol. Bull. 170:145–175.

Maki, J.S., and Mitchell, R. 1985. Involvement of lectins in the settlement and metamorphosis of marine invertebrate larvae. Bull. Mar. Sci. 37:675–683.

Maki, J.S., Rittschof, D., Costlow, J.D., and Mitchell, R. 1988. Inhibition of attachment of larval barnacles, *Balanus amphitrite*, by bacterial surface films. Mar. Biol. 97:199–206.

Maki, J.S., Rittschof, D., Schmidt, A.R., Snyder, A.G., and Mitchell, R. 1989. Factors controlling attachment of bryozoan larvae: a comparison of bacterial films and unfilmed surfaces. Biol. Bull. 177:295–302.

Marsden, J.R. 1987. Coral preference behaviour by planktotrophic larvae of *Spirobranchus giganteus corniculatus* (Serpulidae: Polychaeta). Coral Reefs 6:71–74.

Marsden, J.R., and Meeuwig, J. 1990. Preferences of planktotrophic larvae of the tropical serpulid *Spirobranchus giganteus* (Pallas) for exudates of corals from a Barbados reef. J. Exp. Mar. Biol. Ecol. 137:95–104.

Meadows, P.S., and Campbell, J.I. 1972. Habitat selection by aquatic invertebrates. Adv. Mar. Biol. 10:271–382.

Menge, B.A. 1991. Relative importance of recruitment and other causes of variation in rocky intertidal community structure. J. Exp. Mar. Biol. Ecol. 146:69–100.

Mihm, J.W., Banta, W.C., and Loeb, G.I. 1981. Effects of adsorbed organic and primary fouling films on bryozoan settlement. J. Exp. Mar. Biol. Ecol. 54:167–179.

Miller, R.L., and King, K.R. 1983. Sperm chemotaxis in *Oikopleura dioica* Fol, 1872 (Urochordata: Larvacea). Biol. Bull. 165:419–428.

Mitchell, R., and Kirchman, D. 1984. The microbial ecology of marine surfaces. *In* Marine biodeterioration: an interdisciplinary study, ed. J.D. Costlow and R.C. Tipper, pp. 49–56. Annapolis, Md.: Naval Institute Press.

Moore, P., and Atema, J. 1988. A model of a temporal filter in chemoreception to extract directional information from a turbulent odor plume. Biol. Bull. 174:355–363.

Morse, A.N.C., Froyd, C.A., and Morse, D.E. 1984. Molecules from cyanobacteria and red algae that induce larval settlement and metamorphosis in the mollusc *Haliotis rufescens*. Mar. Biol. 81:293–298.

Morse, A.N.C., and Morse, D.E. 1984a. Recruitment and metamorphosis of *Haliotis* larvae induced by molecules uniquely available at the surfaces of crustose red algae. J. Exp. Mar. Biol. Ecol. 75:191–215.

Morse, A.N.C., and Morse, D.E. 1984b. GABA-mimetic molecules from *Porphyra* (Rhodophyta) induce metamorphosis of *Haliotis* (Gastropoda) larvae. Hydrobiologia 116:155–158.

Morse, D.E. 1985. Neurotransmitter-mimetic inducers of larval settlement and metamorphosis. Bull. Mar. Sci. 37:697–706.

Morse, D.E. 1990. Recent progress in larval settlement: closing the gap between molecular biology and ecology. Bull. Mar. Sci. 46:465–483.

Morse, D.E., Duncan, H., Hooker, N., Baloun, A., and Young, G. 1980a. GABA induces behavioral and developmental metamorphosis in planktonic molluscan larvae. Fed. Proc. 39:3237–3241.

Morse, D.E., Hooker, N., and Duncan, H. 1980b. GABA induces metamorphosis in *Haliotis*. V. Stereochemical specificity. Brain Res. Bull. 5:381–387.

Morse, D.E., Hooker, N., Duncan, H., and Jensen, L. 1979. γ-Aminobutyric acid, a neurotransmitter, induces planktonic abalone larvae to settle and begin metamorphosis. Science 204:407–410.

Morse, D.E., Hooker, N., Morse, A.N.C., and Jensen, R.A. 1988. Control of larval metamorphosis and recruitment in sympatric agariciid corals. J. Exp. Mar. Biol. Ecol. 116:193–217.

Morse, D.E., Tegner, M., Duncan, H., Hooker, N., Trevelyan, G., and Cameron, A. 1980c. Induction of settling and metamorphosis of planktonic molluscan (*Haliotis*) larvae. III. Signalling by metabolites of intact algae is dependent on contact. *In* Chemical signals, ed. D. Müller-Schwarze and R. M. Silverstein, New York: Plenum Press.

Müller, W.A. 1973. Metamorphose-Induktion bei Planulalarven. I. Der bakterielle Induktor. Wilhelm Roux's Arch. Dev. Biol. 173:107–121.

Müller, W.A., and Buchal, G. 1973. Metamorphose-Induktion bei Planulalarven. II. Induktion durch monovalente Kationen: Die Bedeutung des Gibbs-Donnan-Verhältnisses und der Na$^+$/K$^+$-ATPase. Wilhelm Roux's Arch. Dev. Biol. 173:122–135.

Nadeau, L., Paige, J.A., Starczak, V., Capo, T., Lafler, J., and Bidwell, J.P. 1989. Metamorphic competence in *Aplysia californica* Cooper. J. Exp. Mar. Biol. Ecol. 131:171–193.

Neumann, R. 1979. Bacterial induction of settlement and metamorphosis in the planula larvae of *Cassiopea andromeda* (Cnidaria: Scyphozoa, Rhizostomeae). Mar. Ecol. Prog. Ser. 1:21–28.

Nielsen, S.A. 1973. Effect of acetazolamide on larval settlement of *Ostrea lutaria*. Veliger 16:66–67.

Nishihira, M. 1968. Brief experiments on the effect of algal extracts in promoting the settlement of the larvae of *Coryne uchidai* Stechow (Hydrozoa). Bull. Mar. Biol. Stn. Asamushi 13:91–101.

Nott, J.A. 1969. Settlement of barnacle larvae: surface of the antennular attachment disc by scanning electron microscopy. Mar. Biol. 2:248–251.

Nott, J.A., and Foster, B.A. 1969. On the structure of the antennular attachment organ of the cypris larva of *Balanus balanoides* (L.). Philos. Trans. R. Soc. Lond. B 256:115–133.

O'Connell, R.J. 1986. Chemical communication in invertebrates. Experientia 42:232–241.

Oehlschlager, A.C., Pierce, A.M., Pierce, H.D., Jr., and Borden, J.H. 1988. Chemical communication in cucujid grain beetles. J. Chem. Ecol. 14:2071–2098.

Pawlik, J.R. 1986. Chemical induction of larval settlement and metamorphosis in the reef-building tube worm *Phragmatopoma californica* (Polychaeta: Sabellariidae). Mar. Biol. 91:59–68.

Pawlik, J.R. 1987. *Bocquetia rosea*, new genus, new species, an unusual rhizocephalan parasite of a sponge-inhabiting barnacle, *Membranobalanus orcutti* (Pilsbry), from California. J. Crustacean Biol. 7:265–273.

Pawlik, J.R. 1988a. Larval settlement and metamorphosis of two gregarious sabellariid polychaetes: *Sabellaria alveolata* compared with *Phragmatopoma californica*. J. Mar. Biol. Assoc. U.K. 68:101–124.

Pawlik, J.R. 1988b. Larval settlement and metamorphosis of sabellariid polychaetes, with special reference to *Phragmatopoma lapidosa*, a reef-building species, and *Sabellaria floridensis*, a non-gregarious species. Bull. Mar. Sci. 43:41–60.

Pawlik, J.R. 1989. Larvae of the sea hare *Aplysia californica* settle and metamorphose on an assortment of macroalgal species. Mar. Ecol. Prog. Ser. 51:195–199.

Pawlik, J.R. 1990. Natural and artificial induction of metamorphosis of *Phragmatopoma lapidosa californica* (Polychaeta: Sabellariidae), with a critical look at the effects of bioactive compounds on marine invertebrate larvae. Bull. Mar. Sci. 46:512–536.

Pawlik, J.R. 1992. Chemical ecology of the settlement of benthic marine invertebrates. Oceanogr. Mar. Biol. Annu. Rev., in press.

Pawlik, J.R., Butman, C.A., and Starczak, V.R. 1991. Hydrodynamic facilitation of gregarious settlement of a reef-building tube worm. Science 251:421–424.

Pawlik, J.R., and Chia, F.S. 1991. Larval settlement of *Sabellaria cementarium* Moore, and comparisons with other species of sabellariid polychaetes. Can. J. Zool. 69:765–770.

Pawlik, J.R., and Faulkner, D.J. 1986. Specific free fatty acids induce larval settlement and metamorphosis of the reef-building tube worm *Phragmatopoma californica* (Fewkes). J. Exp. Mar. Biol. Ecol. 102:301–310.

Pawlik, J.R., and Faulkner, D.J. 1988. The gregarious settlement of sabellariid polychaetes: new perspectives on chemical cues. *In* Marine biodeterioration. Advanced techniques applicable to the Indian Ocean, ed. M.F. Thompson, R. Sarojini, and R. Nagabhushanam, pp. 475–487. New Delhi: Oxford and IBH Publishing.

Pearce, C.M., and Scheibling, R.E. 1990. Induction of settlement and metamorphosis in the sand dollar *Echinarachnius parma*: evidence for an adult-associated factor. Mar. Biol. 107:363–369.

Pechenik, J.A., and Heyman, W.D. 1987. Using KCl to determine size at competence for larvae of the marine gastropod *Crepidula fornicata* (L.). J. Exp. Mar. Biol. Ecol. 112:27–38.

Pelter, A., Ballantine, J.A., Murray-Rust, P., Ferrito, V., and Psaila, A.F. 1978. The structures of anhydrobonellin and bonellin, the physiologically active pigment from the marine echiuroid *Bonellia viridis*. Tetrahedron Lett. 21:1881–1884.

Pennington, J.T. 1985. The ecology of fertilization of echinoid eggs: the consequences of sperm dilution, adult aggregation, and synchronous spawning. Biol. Bull. 169:417–430.

Pennington, J.T., and Hadfield, M.G. 1989. Larvae of a nudibranch mollusc (*Phestilla sibogae*) metamorphose when exposed to common organic solvents. Biol. Bull. 177:350–355.

Petersen, C.G.J. 1913. Valuation of the sea. II. The animal communities of the sea-bottom and their importance for marine zoogeography. Rep. Danish Biol. Stn. 21:1–44.

Pilger, J. 1978. Settlement and metamorphosis in the Echiura: a review. *In* Settlement and metamorphosis of marine invertebrate larvae, ed. F.S. Chia and M.E. Rice, pp. 103–112. New York: Elsevier.

Pires, A., and Hadfield, M.G. 1991. Oxidative breakdown products of catecholamines and hydrogen peroxide induce partial metamorphosis in the nudibranch *Phestilla sibogae* Bergh (Gastropoda: Opisthobranchia). Biol. Bull. 180:310–317.

Prestwich, G.D. 1985. Communication in insects. II. Molecular communication of insects. Q. Rev. Biol. 60:437–456.

Prytherch, H.F. 1931. The role of copper in the setting and metamorphosis of the oyster. Science 73:429–431.

Raimondi, P.T. 1988a. Rock type affects settlement, recruitment, and zonation of the barnacle *Chthamalus anisopoma* Pilsbry. J. Exp. Mar. Biol. Ecol. 123:253–267.

Raimondi, P.T. 1988b. Settlement cues and determination of the vertical limit of an intertidal barnacle. Ecology 69:400–407.

Raimondi, P.T. 1991. Settlement behavior of *Chthamalus anisopoma* larvae largely determines the adult distribution. Oecologia 85:349–360.

Reed, C.G. 1988a. Organization of the nervous system and sensory organs in the larva of the marine bryozoan *Bowerbankia gracilis* (Ctenostomata: Vesiculariidae): functional significance of the apical disk and pyriform organ. Acta Zool. 69:177–194.

Reed, C.G. 1988b. Organization and isolation of the ciliary locomotory and sensory organs of marine bryozoan larvae. *In* Marine biodeterioration. Advanced techniques applicable to the Indian Ocean, ed. M.F. Thompson, R. Sarojini, and R. Nagabhushanam, pp. 397–408. New Delhi: Oxford and IBH Publishing.

Reed, C.G., Ninos, J.M., and Woollacott, R.M. 1988. Bryozoan larvae as mosaics of multifunctional ciliary fields: ultrastructure of the sensory organs of *Bugula stolonifera* (Cheilostomata: Cellularioidea). J. Morphol. 197:127–145.

Rice, M.E. 1988. Factors influencing larval metamorphosis in *Golfingia misakiana* (Sipuncula). Bull. Mar. Sci. 39:362–375.

Rittschof, D., Maki, J., Mitchell, R., and Costlow, J.D. 1986. Ion and neuropharmacological studies of barnacle settlement. Neth. J. Sea Res. 20:269–275.

Rittschof, D., Williams, L.G., Brown, B., and Carriker, M.R. 1983. Chemical attraction of newly hatched oyster drills. Biol. Bull. 164:493–505.

Roininen, H., and Tahvanainen, J. 1989. Host selection and larval performance of two willow-feeding sawflies. Ecology 70:129–136.

Rowley, R.J. 1989. Settlement and recruitment of sea urchins (*Strongylocentrotus* spp.) in a sea-urchin barren ground and a kelp bed: are populations regulated by settlement or post-settlement processes? Mar. Biol. 100:485–494.

Ryland, J.S. 1974. Behaviour, settlement and metamorphosis of bryozoan larvae: a review. Thalassia Jugosl. 10:239–262.

Satterlie, R.A., and Cameron, R.A. 1985. Electrical activity at metamorphosis in the larvae of the sea urchin *Lytechinus pictus* (Echinoidea: Echinodermata). J. Exp. Zool. 235:197–204.

Scheltema, R.S. 1961. Metamorphosis of the veliger larvae of *Nassarius obsoletus* (Gastropoda) in response to bottom sediment. Biol. Bull. 120:92–109.

Scheltema, R.S. 1974. Biological interactions determining larval settlement of marine invertebrates. Thalassia Jugosl. 10:263–269.

Scheltema, R.S., Williams, I.P., Shaw, M.A., and Loudon, C. 1981. Gregarious settlement by the larvae of *Hydroides dianthus* (Polychaeta: Serpulidae). Mar. Ecol. Prog. Ser. 5:69–74.

Sebens, K.P. 1983. The larval and juvenile ecology of the temperate octocoral *Alcyonium siderium* Verrill. I. Substratum selection by benthic larvae. J. Exp. Mar. Biol. Ecol. 71:73–89.

Shimada, I. 1978. The stimulating effect of fatty acids and amino acid derivatives on the labellar sugar receptor of the fleshfly. J. Gen. Physiol. 71:19–36.

Slattery, M. 1987. Settlement and metamorphosis of red abalone (*Haliotis rufescens*) larvae: a critical examination of mucous, diatoms, and γ-aminobutyric acid (GABA) as inductive substrates. M.A. thesis, San Jose State University, California.

Smith, C.M., and Hackney, C.T. 1989. The effects of hydrocarbons on the setting of the American oyster, *Crassostrea virginica*, in intertidal habitats in southeastern North Carolina. Estuaries 12:42–48.

Spindler, K.D., and Müller, W.A. 1972. Induction of metamorphosis by bacteria and a lithium-pulse in the larvae of *Hydractinia echinata* (Hydrozoa). Wilhelm Roux's Arch. Dev. Biol. 169:271–280.

Städler, E. 1984. Contact chemoreception. *In* Chemical ecology of insects, ed. W.J. Bell and R.T. Cardé, pp. 3–35. Sunderland, Mass.: Sinauer Associates.

Stevens, P.M. 1990. Specificity of host recognition of individuals from different host races of symbiotic pea crabs (Decapoda: Pinnotheridae). J. Exp. Mar. Biol. Ecol. 143:193–207.

Strathmann, M. 1987. Reproduction and development of marine invertebrates of the northern Pacific coast. Seattle: University of Washington Press.

Strathmann, R.R., and Branscomb, E.S. 1979. Adequacy of cues to favorable sites used by settling larvae of two intertidal barnacles. *In* Reproductive ecology of marine invertebrates, ed. S.E. Stancyk, pp. 77–89. Columbia: University of South Carolina Press.

Strathmann, R.R., Branscomb, E.S., and Vedder, K. 1981. Fatal errors in set as a cost of dispersal and the influence of intertidal flora on set of barnacles. Oecologia 48:13–18.

Suer, A.L., and Phillips, D.W. 1983. Rapid, gregarious settlement of the larvae of the marine echiuran *Urechis caupo* Fisher & MacGinitie 1928. J. Exp. Mar. Biol. Ecol. 67:243–259.

Sulkin, S.D. 1984. Behavioral basis of depth regulation in the larvae of brachyuran crabs. Mar. Ecol. Prog. Ser. 15:181–205.

Sutherland, J.P. 1990. Recruitment regulates demographic variation in a tropical intertidal barnacle. Ecology 71:955–972.

Svane, I., Havenhand, J.N., and Jørgensen, A.J. 1987. Effects of tissue extract of adults on metamorphosis in *Ascidia mentula* O.F. Müller and *Ascidiella scabra* (O.F. Müller). J. Exp. Mar. Biol. Ecol. 110:171–181.

Sweatman, H. 1988. Field evidence that settling coral reef fish larvae detect resident fishes using dissolved chemical cues. J. Exp. Mar. Biol. Ecol. 124:163–174.

Switzer-Dunlap, M. 1978. Larval biology and metamorphosis of aplysiid gastropods. *In* Settlement and metamorphosis of marine invertebrate larvae, ed. F.S. Chia and M.E. Rice, pp. 197–206. New York: Elsevier.

Thompson, T.E. 1958. The natural history, embryology, larval biology, and postlarval development of *Adalaria proxima* (Gastropoda: Opisthobranchia). Philos. Trans. R. Soc. Lond. B 242:1–58.

Thorson, G. 1964. Light as an ecological factor in the dispersal and settlement of larvae of marine bottom invertebrates. Ophelia 1:167–208.

Thorson, G. 1966. Some factors influencing the recruitment and establishment of marine benthic communities. Neth. J. Sea Res. 3:267–293.

Todd, C.D., Bentley, M.G., and Havenhand, J.N. 1991. Larval metamorphosis of the opisthobranch mollusc *Adalaria proxima* (Gastropoda: Nudibranchia): the effects of choline and elevated potassium ion concentration. J. Mar. Biol. Assoc. U.K. 71:53–72.

Torrence, S.A., and Cloney, R.R. 1988. Larval sensory organs of ascidians. *In* Marine biodeterioration. Advanced techniques applicable to the Indian Ocean, ed. M.F. Thompson, R. Sarojini, and R. Nagabhushanam, pp. 151–163. New Delhi: Oxford and IBH Publishing.

Trapido-Rosenthal, H.G., and Morse, D.E. 1986. Availability of chemosensory receptors is down-regulated by habituation of larvae to a morphogenetic signal. Proc. Natl. Acad. Sci. USA 83:7658–7662.

Vandermeulen, J.H. 1974. Studies on reef corals. II. Fine structure of planktonic planula larva of *Pocillopora damicornis*, with emphasis on the aboral epidermis. Mar. Biol. 27:239–249.

VanDover, C.L., Berg, C.J., and Turner, R.D. 1988. Recruitment of marine invertebrates to hard substrates at deep-sea hydrothermal vents on the East Pacific Rise and Galapagos spreading center. Deep-Sea Res. 35:1833–1849.

Veitch F.P., and Hidu, H. 1971. Gregarious setting in the American oyster *Crassostrea virginica* Gmelin. I. Properties of a partially purified "setting factor." Chesapeake Sci. 12:173–178.

Vinnikov, Y.A. 1982. Chemoreceptor cells (olfactory and taste cells). *In* Evolution of receptor cells. Molecular biology, biochemistry and biophysics, no. 34, pp. 29–58. New York: Springer Verlag.

Vinson, S.B. 1984. Parasitoid-host relationship. *In* Chemical ecology of insects, ed. W.J. Bell and R.T. Cardé, pp. 205–233. Sunderland, Mass.: Sinauer Associates.

Walker, G., and Yule, A.B. 1984. Temporary adhesion of the barnacle cyprid: the existence of an antennular adhesive secretion. J. Mar. Biol. Assoc. U.K. 64:679–686.

Walters, L.J., and Wethey, D.S. 1991. Settlement, refuges, and adult body form in colonial marine invertebrates: a field experiment. Biol. Bull. 180:112–118.

Watanabe, J.M. 1984. The influence of recruitment, competition, and benthic predation on spatial distributions of three species of kelp forest gastropods (Trochidae: *Tegula*). Ecology 65:920–936.

Weiner, R.M., Segall, A.M., and Colwell, R.R. 1985. Characterization of a marine bacterium associated with *Crassostrea virginica* (the eastern oyster). Appl. Environ. Microbiol. 49:83–90.

Williams, G.B. 1964. The effect of extracts of *Fucus serratus* in promoting the settlement of larvae of *Spirorbis borealis* (Polychaeta). J. Mar. Biol. Assoc. U.K. 44:397–414.

Wilson, D.P. 1968. The settlement behaviour of the larvae of *Sabellaria alveolata* (L.). J. Mar. Biol. Assoc. U.K. 48:387–435.

Wilson, D.P. 1974. *Sabellaria* colonies at Duckpool, North Cornwall, 1971–1972, with a note for May 1973. J. Mar. Biol. Assoc. U.K. 54:393–436.

Wodicka, L.M., and Morse, D.E. 1991. cDNA sequences reveal mRNAs for two Gα signal transducing proteins from larval cilia. Biol. Bull. 180:318–327.

Yool, A.J., Grau, S.M., Hadfield, M.G., Jensen, R.A., Markell, D.A., and Morse, D.E. 1986. Excess potassium induces larval metamorphosis of four marine invertebrate species. Biol. Bull. 170:255–266.

Young, C.M., and Chia, F.S. 1987. Abundance and distribution of pelagic larvae as influenced by predation, behavior, and hydrographic factors. *In* Reproduction of marine invertebrates, vol. 9, ed. A.C. Giese, J.S. Pearse, and V.B. Pearse, pp. 385–463. Palo Alto, Calif.: Blackwell.

Yule, A.B., and Crisp, D.J. 1983. Adhesion of cypris larvae of the barnacle, *Balanus balanoides*, to clean and arthropodin treated surfaces. J. Mar. Biol. Assoc. U.K. 63:261–271.

Yule, A.B., and Walker, G. 1984. The temporary adhesion of barnacle cyprids: effects of some differing surface characteristics. J. Mar. Biol. Assoc. U.K. 64:429–439.

Yule, A.B., and Walker, G. 1987. Adhesion in barnacles. *In* Barnacle biology, ed. A.J. Southward, pp. 389–402. Rotterdam: A.A. Balkema.

Yvin, J.C., Chevolot, L., Chevolot-Magueur, A.M., and Cochard, J.C. 1985. First isola-
    tion of jacaranone from an alga, *Delesseria sanguinea*. A metamorphosis inducer
    of *Pecten* larvae. J. Nat. Prod. 48:814–816.
Zann, L.P. 1980. Living together in the sea. Neptune, N.J.: T.F.H. Publications.
Zobell, C.E., and Allen, E.C. 1935. The significance of marine bacteria in the fouling
    of submerged surfaces. J. Bacteriol. 29:239–251.

Index